The Algebra of Calculus

with Trigonometry and Analytic Geometry

Eric Braude

D. C. Heath and Company

Lexington, Massachusetts   Toronto

Dedicated to the memory of
Fred Braude
Loving father, gentle man

Published simultaneously in Canada.

Printed in the United States of America.

International Standard Book Number: 0-669-21885-5

6   7   8   9

Preface

For many students, a calculus course represents the first time they are required to bring into play assorted topics from algebra, trigonometry, and geometry. These students are required to recall these topics in the course of solving larger problems from calculus itself. This is a much more difficult process than studying one topic in a textbook, and then answering the test questions on that topic. Even the best students are often stumped by the need to recall background topics on demand. In fact, most students do not even have a reference text in their possession at the time they enter college. As its name suggests, *The Algebra of Calculus* is designed to fill this need by reviewing and reinforcing the skills from algebra, trigonometry, and analytic geometry required in calculus.

The Algebra of Calculus serves the following purposes.

- The text provides a **quick review** of all the types of problems using algebra that calculus students *must* be able to solve.
- The text provides a **reference** for the algebraic skills required as the student works through the calculus course or in parallel with a precalculus course.
- The text **alleviates the apprehension** many students experience in approaching the calculus course by reinforcing their previously learned math skills.

Instructors, students, and tutors will find that this book emphasizes the reinforcement of algebra skills and the integration of these skills into the broader context of solving calculus problems.

Each chapter of *The Algebra of Calculus* begins with a review of the topic and progresses through solved examples, similar problems with answers, and finally, exercises. This step-by-step skill development is enhanced by the following features.

- In the **Review of Fundamentals,** brief and precise descriptions are given for several algebraic procedures. This review appears at the beginning of each chapter.
- Over **200 solved examples** illustrate how algebra, trigonometry, and analytic geometry are used in calculus.
- **Similar problems** with answers appear after 2–3 examples of a particular type of problem. They offer students practice in using the skills just illustrated in the solved example and provide immediate feedback.
- A set of **Exercises** concludes each chapter. These exercises give students additional opportunity to practice the algebra skills studied in the chapter. Answers to all exercises are given in the back of the book.
- The book is **keyed** to the following texts via reference charts in the front of the book.

  Calculus with Analytic Geometry, 4/E (Larson/Hostetler/Edwards)

  Calculus with Analytic Geometry, Alternate Edition, 4/E (Larson/Hostetler)

  Precalculus, 2/E (Larson/Hostetler)

  Brief Calculus with Applications, 3/E (Larson/Hostetler/Edwards)

  For students in engineering calculus, business calculus, and precalculus who are not using the texts listed, there is also a chart organized by course topic.
- A set of **pretests** allows the student to assess his or her skill level to determine which types of problems need additional practice. Located in the Appendix, these pretests consist of the examples that are solved in each chapter so that students can obtain feedback on test questions they are unable to answer.

Throughout the time that I spent writing this book, my wife Sherry kept faith that my project was an increasingly important one for calculus students. I thank Sherry here for this support. I would also like to thank Ron Larson for his encouragement, Nancy Stout and Linda Bollinger for typing the manuscript, Richard Bambauer for preparing the art, Paula Sibeto for proof-reading the manuscript, Ed Quigly for his early recognition of the value of this work, and Ann Marie Jones, Cathy Cantin, and Carolyn Johnson, of D. C. Heath who have made editorial collaboration a pleasure.

Eric J. Braude

To The Student

If you are taking engineering calculus, business calculus, or precalculus, *The Algebra of Calculus* will help you master the algebra skills you must have to work calculus problems. Numerous reviews, solved examples, similar problems with answers, exercises (with answers at the back of the book), and pretests give practice using algebra, trigonometry, and analytic geometry for all types of calculus problems.

To make it easy for you to use this book, it is **keyed** to four textbooks:

Calculus with Analytic Geometry, 4/E (Larson/Hostetler/Edwards)

Calculus with Analytic Geometry, Alternate Edition, 4/E (Larson/Hostetler)

Precalculus, 2/E (Larson/Hostetler)

Brief Calculus with Applications, 3/E (Larson/Hostetler/Edwards)

If you are using a different book in your course, this text is keyed to your book by topic. This information is given in the Reference Charts which begin on page ix.

Use the Reference Charts to determine the material in *The Algebra of Calculus* that will help you with the topic you are currently covering in class. Look up your text section number (or topic) on the left-hand side of the chart. The right-hand side of the chart will show page numbers in *The Algebra of Calculus* for the extra help that you need. The general **index** in the back of the text will also serve this purpose.

When you have chosen the appropriate chapter in *The Algebra of Calculus*, begin by reading the **Review of Fundamentals**. (This section also will serve as a convenient review later, before a class test.) Then, study the **examples** and their solutions. Next, work the **similar problems** that follow the examples and check your answers. The method for solving each of the similar problems is the same as the one used in the preceding examples. When you feel confident that you understand the method for solving a particular type of problem, go on to the next examples. Finally, work the exercises and check your answers in the back of the book. Alternatively, you may want to go directly to examples, problems, or exercises.

If you are about to start a new topic in your calculus course, you will find the **pretests** helpful. (The questions in the pretests are the examples that are solved in each chapter.) Take the test that pertains to the calculus topic you are studying. If you are able to answer all the questions correctly, you can feel confident that you know how to use all the algebra you will need to solve the calculus problems in that section. If you answer some problems incorrectly, work through that chapter of *The Algebra of Calculus* for extra practice. The complete solutions to the questions on the pretest are found in each chapter. If you are currently taking a precalculus course, the pretests will help you determine how well you are prepared for the calculus course.

Contents

Cross Reference Charts vi

Chapter   1   Selected Elementary Algebra 1

Chapter   2   Factoring 8

Chapter   3   Equations 13

Chapter   4   Slope 25

Chapter   5   Common Graphs and Their Equations 30

Chapter   6   Inequalities 45

Chapter   7   Functions I: Notation, Domain, Range, and Simpler Graphs . 56

Chapter   8   Functions II: Combinations of Functions, Difficult Graphs . . 67

Chapter   9   Simultaneous Equations 79

Chapter 10   Completing the Square 84

Chapter 11   Exponents and Radicals 88

Chapter 12   Length, Area, and Volume Formulas 96

Chapter 13   Logarithms 104

Chapter 14   Trigonometry I: Trigonometric Functions and Their Inverses 109

Chapter 15   Trigonometry II: Trigonometric Identities 117

Chapter 16   Sigma Notation, The Binomial Theorem, and Mathematical Induction 127

Appendix: Pretests 133

Pretest Answers 151

Answers to Chapter Exercises 163

Index 175

Cross Reference Chart for Calculus, 4E
Larson/Hostetler/Edwards, 1990

Calculus, 4th Edition (Section Numbers)	The Algebra of Calculus (Page Numbers)
Chapter 1 **The Cartesian Plane and Functions**	
1.1 Real Numbers and the Real Line	45–52
1.2 The Cartesian Plane	14, 30, 41, 86
1.3 Graphs of Equations	8, 9, 13, 15, 16, 23, 24, 30–42
1.4 Lines in the Plane	25–27, 30, 33, 35–37
1.5 Functions	2, 30, 41, 56, 57, 60–63, 67–69
1.6 Review of Trigonometric Functions	30, 31, 109–112, 117
Chapter 2 **Limits and Their Properties**	
2.1 An Introduction to Limits	45, 46, 51–56, 63, 88, 109, 110, 112
2.2 Properties of Limits	1, 6, 8, 11, 88, 109–111
2.3 Techniques for Evaluating Limits	8, 11, 92, 111, 117
2.4 Continuity and One-Sided Limits	56, 57, 63, 88, 109, 110
2.5 Infinite Limits	8, 11, 56, 67, 117
Chapter 3 **Differentiation**	
3.1 The Derivative and the Tangent Line Problem	1–6, 8, 51, 60, 61, 88, 91
3.2 Velocity, Acceleration, and Other Rates of Change	1, 5, 30–32
3.3 Differentiation Rules for Powers, Constant Multiples, Sums, Sines, and Cosines	13–15, 88–90, 109–111
3.4 Differentiation Rules for Products, Quotients, Secants, and Tangents	4, 8, 11, 88, 92, 114, 117
3.5 The Chain Rule	6, 10, 19, 69, 70, 88–92, 113, 114
3.6 Implicit Differentiation	1, 4, 14, 27, 30, 52, 56, 57, 67, 75
3.7 Related Rates	96, 99, 100, 109
Chapter 4 **Applications of Differentiation**	
4.1 Extrema on an Interval	8, 11, 51, 90, 109–111
4.2 Rolle's Theorem and the Mean Value Theorem	13–15, 25, 26
4.3 Increasing and Decreasing Functions and the First Derivative Test	13, 21, 88, 109–112
4.4 Concavity and the Second Derivative Test	8–11
4.5 Limits at Infinity	1, 4–6, 8, 11, 109, 110
4.6 A Summary of Curve Sketching	8, 10, 11, 13, 20, 21, 88, 90, 109–111, 117
4.7 Optimization Problems	13, 18–21, 97–99, 111
4.8 Newton's Method	88–91, 118, 119
4.9 Differentials	45, 46, 52, 88, 89, 96, 109–111
4.10 Business & Economics Applications	

Calculus, 4th Edition (Section Numbers)		The Algebra of Calculus (Page Numbers)
Chapter 5	**Integration**	
5.1	Antiderivatives and Indefinite Integration	89–92, 117
5.2	Area	127–131
5.3	Riemann Sums and the Definite Integral	127–130
5.4	The Fundamental Theorem of Calculus	63, 79, 88, 109–112
5.5	Integration by Substitution	88, 89, 114, 117
5.6	Numerical Integration	109–112
Chapter 6	**Logarithmic, Exponential, & Other Transcendental Functions**	
6.1	The Natural Logarithmic Function and Differentiation	104–107, 117
6.2	The Natural Logarithmic Function and Integration	104, 117
6.3	Inverse Functions	25, 69–74
6.4	Exponential Functions: Differentiation and Integration	88, 94, 104
6.5	Bases Other Than e and Applications	104–107
6.6	Growth and Decay	94, 104
6.7	Inverse Trigonometric Functions and Differentiation	109–115
6.8	Inverse Trigonometric Functions: Integration and Completing the Square	84–86, 91
6.9	Hyperbolic Functions	88, 104
Chapter 7	**Applications of Integration**	
7.1	Area of a Region Between Two Curves	13, 18, 30, 38–40, 45, 80, 88, 117
7.2	Volume: The Disc Method	30–42, 45, 46, 52, 92, 96, 100, 110
7.3	Volume: The Shell Method	1, 2, 88–90, 96–98
7.4	Arc Length and Surfaces of Revolution	1, 88–91, 117
7.5	Work	96–98
7.6	Fluid Pressure and Fluid Force	25, 26, 30, 39, 96
7.7	Moments, Centers of Mass, and Centroids	30, 39, 79, 80, 82, 96–98, 127
Chapter 8	**Integration Techniques, L'Hôpital's Rule, and Improper Integrals**	
8.1	Basic Integration Formulas	88–90, 109, 117
8.2	Integration by Parts	109, 110, 114, 117–120
8.3	Trigonometric Integrals	84, 88, 109–112, 114, 117–122
8.4	Trigonometric Substitution	84, 96, 97, 109, 110, 117
8.5	Partial Fractions	79, 81, 104, 107
8.6	Integration by Tables and Other Integration Techniques	91, 104, 107, 109, 110, 117
8.7	Indeterminate Forms and L'Hôpital's Rule	4, 5, 90, 93, 105
8.8	Improper Integrals	30, 91, 96, 104, 109, 110, 112

Cross Reference Chart for Calculus, Alternate 4E
Larson/Hostetler, 1990

Calculus, Alternate 4th Edition (Section Numbers)		The Algebra of Calculus (Page Numbers)
Chapter 1	**The Cartesian Plane and Functions**	
1.1	Real Numbers and the Real Line	45–52
1.2	The Cartesian Plane	14, 30, 41, 86
1.3	Graphs of Equations	8, 9, 13, 15, 16, 23, 24, 30–42
1.4	Lines in the Plane	25–27, 30, 33, 35–37
1.5	Functions	2, 30, 41, 56, 57, 60–63, 67–69
Chapter 2	**Limits and Their Properties**	
2.1	An Introduction to Limits	63
2.2	Techniques for Evaluating Limits	8, 11, 92
2.3	Continuity	56, 57, 63, 88
2.4	Infinite Limits	8, 11, 56, 67
2.5	ϵ–δ Definition of Limits	45, 46, 51–56, 76, 77
Chapter 3	**Differentiation**	
3.1	The Derivative and the Tangent Line Problem	1–6, 8, 51, 60, 61, 88, 91
3.2	Velocity, Acceleration, and Other Rates of Change	1, 5, 30–32
3.3	Differentiation Rules for Powers, Constant Multiples, and Sums	13–15, 88–90
3.4	Differentiation Rules for Products and Quotients	4, 8, 11, 88, 92
3.5	The Chain Rule	6, 10, 19, 69, 70, 88–92
3.6	Implicit Differentiation	1, 4, 14, 27, 30, 52, 56, 57, 67, 75
3.7	Related Rates	96, 99, 100
Chapter 4	**Applications of Differentiation**	
4.1	Extrema on an Interval	8, 11, 51, 90
4.2	Rolle's Theorem and the Mean Value Theorem	13–15, 25, 26
4.3	Increasing and Decreasing Functions and the First Derivative Test	13, 21, 88
4.4	Concavity and the Second Derivative Test	8–11
4.5	Limits at Infinity	1, 4–6, 8, 11
4.6	A Summary of Curve Sketching	8, 10, 11, 13, 20, 21, 88, 90
4.7	Optimization Problems	13, 18–21, 97–99
4.8	Newton's Method	88–91
4.9	Differentials	45, 46, 52, 88, 89, 96
4.10	Business & Economics Applications	

		Calculus, Alternate 4th Edition (Section Numbers)	**The Algebra of Calculus** (Page Numbers)

Chapter 5   Integration

5.1	Antiderivatives and Indefinite Integration	89–92
5.2	Area	127–131
5.3	Riemann Sums and the Definite Integral	127–130
5.4	The Fundamental Theorem of Calculus	63, 79, 88
5.5	Integration by Substitution	88, 89
5.6	Numerical Integration	

Chapter 6   Applications of Integration

6.1	Area of a Region Between Two Curves	13, 18, 30, 38–40, 45, 80, 88
6.2	Volume: The Disc Method	30–42, 45, 46, 52, 92, 96, 100
6.3	Volume: The Shell Method	1, 2, 88–90, 96–98
6.4	Arc Length and Surfaces of Revolution	1, 88–91
6.5	Work	96–98
6.6	Fluid Pressure and Fluid Force	25, 26, 30, 39, 96
6.7	Moments, Centers of Mass, and Centroids	30, 39, 79, 80, 82, 96–98, 127

Chapter 7   Exponential and Logarithmic Functions

7.1	Exponential Functions	88
7.2	Differentiation and Integration of Exponential Functions	88, 94, 104
7.3	Inverse Functions	25, 69–74
7.4	Logarithmic Functions	104–107
7.5	Logarithmic Functions and Differentiation	104–107
7.6	Logarithmic Functions and Integration	104–107
7.7	Growth and Decay	94, 104
7.8	Indeterminate Forms and L'Hôpital's Rule	4, 5, 90, 93, 105

Chapter 8   Trigonometric Functions and Inverse Trigonometric Functions

8.1	Review of Trigonometric Functions	30, 31, 109, 112, 117
8.2	Graphs and Limits of Trigonometric Functions	109–111
8.3	Derivatives of Trigonometric Functions	109–111, 113, 114, 117
8.4	Integrals of Trigonometric Functions	109–112, 114, 117
8.5	Inverse Trigonometric Functions and Differentiation	109–115
8.6	Inverse Trigonometric Functions: Integration and Completing the Square	84–86, 91
8.7	Hyperbolic Functions	88, 104

Chapter 9   Integration Techniques, L'Hôpital's Rule, and Improper Integrals

9.1	Basic Integration Formulas	88–90, 109, 117
9.2	Integration by Parts	109, 110, 114, 117–120
9.3	Trigonometric Integrals	84, 88, 109–112, 114, 117–122
9.4	Trigonometric Substitution	84, 96, 97, 109, 110, 117
9.5	Partial Fractions	79, 81, 104, 107
9.6	Integration by Tables and Other Integration Techniques	91, 104, 107, 109, 110, 117
9.7	Improper Integrals	30, 91, 96, 104, 109, 110, 112

Cross Reference Chart for General Calculus Topics

Calculus Topic	The Algebra of Calculus (Page Numbers)
Analytic geometry	
elementary	25–27, 31–37
other	38–41, 56, 57, 71, 82
Antidifferentiation	See "Integration"
Applications	
of differentiation	13–21, 96–101
of integration	13, 15–18, 79, 82, 96–98
Applied maxima and minima problems	13, 15–18, 56, 61, 96–101
Areas	
between curves	13, 15–18, 56, 65, 79, 82
of surfaces of revolution	65, 96–98
under a curve	13, 15–18, 65, 90, 104, 111, 112
using summation	127–131
Chain Rule	67, 69, 113
Circle, equation of	31, 41, 86
Composite functions	56, 67, 69, 70, 113
Concavity	45–49
Continuity	45, 46, 51–53
Curve(s)	
sketching	30, 31, 35–41, 56–59, 63, 64, 71, 74, 75
slope of	25–27, 30, 33, 37
Derivative(s)	
by Chain Rule	67, 69, 70, 113
implicit	13, 14
logarithmic	104, 106, 107
of trigonometric functions	25–27, 56, 60, 113, 114
theory and proofs	67, 68, 127
using formulas (Power, Product, etc.)	11, 67–70, 88, 89, 91
using limits	1–11, 56, 60
Differentiation	See "Derivative"
Equation of tangents	25–27, 30, 33, 34
Functions	56–65, 67–77, 93, 100, 101, 105
Graphs	See "Curves"
of functions	56–58, 62–64, 67–76, 93, 105, 109, 111
Implicit Differentiation	13, 14

Calculus Topic	The Algebra of Calculus (Page Numbers)
Increasing functions	41–50, 56, 61
Infinite series	127, 128
Integration	
by partial fractions	79, 81
definite	13, 15–18, 65, 90, 104, 111, 112
elementary	88–92
involving quadratic expressions	84, 85
of trigonometric functions	109–115, 117, 121–124
theory and proofs	41, 109, 110, 114, 115, 127–130
Inverse functions	67, 69, 70, 109, 110, 114, 115
inverse trigonometric functions	67, 69, 70, 109, 110, 114, 115
Limits	
evaluation of	1–8, 11
theory of	41, 42, 51–53, 56, 60, 62, 76
Logarithms	104–107
differentiation	104, 106, 107
elementary	104
in integration	104, 106, 107
Maxima and minima	13, 15–20, 56, 62, 97–101
Normal lines	25, 26, 33
Partial fractions	79, 81, 104, 106, 107
Polar coordinates	109–112, 117, 121–125
Power Rule	88, 89, 90, 92
Related rates	14
Series	127–131
Sigma notation	127–131
Slope	25–27
Solids of revolution	96–98, 127–129
Tangent lines	25–27, 30, 33, 34
Trigonometric functions	109–125
integrals of	109–115, 117, 121–124
inverse	67, 72, 75, 109, 110, 114, 115
substitutions	84, 109, 110, 117, 121

Cross Reference Chart for Precalculus, 2E
Larson/Hostetler, 1989

	Precalculus, 2nd Edition (Section Numbers)	The Algebra of Calculus (Page Numbers)
Chapter 1	**Review of Basic Algebra**	
1.1	The Real Number System	1–5, 45–53
1.2	Exponents and Radicals	6, 88–95
1.3	Polynomials: Special Products and Factoring	1, 8–12
1.4	Fractional Expressions	1–14, 88, 90
1.5	Linear Equations and Quadratic Equations	13–24, 84–87, 97
1.6	Inequalities	45–55
1.7	Algebraic Errors and Some Algebra of Calculus	1–12, 88–92
Chapter 2	**Functions and Graphs**	
2.1	The Cartesian Plane	
2.2	Graphs of Equations	30–44
2.3	Lines in the Plane	25–29, 30–37
2.4	Functions	56–65, 82, 97–99
2.5	Graphs of Functions	56–65, 68
2.6	Combinations of Functions	56, 67–77
2.7	Inverse Functions	30, 67, 72, 73
2.8	Variation and Mathematical Models	
Chapter 3	**Polynomial and Rational Functions**	
3.1	Quadratic Functions	30, 31, 38–40
3.2	Polynomial Functions of Higher Degree	13–18, 30, 31, 56
3.3	Polynomial Division and Synthetic Division	18
3.4	Real Zeros of Polynomial Functions	13, 18, 97
3.5	Complex Numbers	
3.6	Complex Zeros and the Fundamental Theorem of Algebra	
3.7	Rational Functions and Their Graphs	8–10, 17, 18, 56
3.8	Partial Fractions	79, 81
Chapter 4	**Exponential and Logarithmic Functions**	
4.1	Exponential Functions	88, 93, 94
4.2	Logarithmic Functions	72, 73, 104–106
4.3	Properties of Logarithms	104–107
4.4	Solving Exponential and Logarithmic Equations	88, 104–107
4.5	Applications	

Precalculus, 2nd Edition (Section Numbers)		The Algebra of Calculus (Page Numbers)
Chapter 5	**Trigonometry**	
5.1	Radian and Degree Measure	109
5.2	The Trigonometric Functions and the Unit Circle	109–112
5.3	Trigonometric Functions of an Acute Angle	99, 109–111, 117
5.4	Trigonometric Functions of Any Angle	109–112, 114, 117
5.5	Graphs of Sine and Cosine	109, 110
5.6	Graphs of Other Trigonometric Functions	
5.7	Additional Graphing Techniques	
5.8	Inverse Trigonometric Functions	109–111, 114, 115
5.9	Applications of Trigonometry	
Chapter 6	**Analytic Trigonometry**	
6.1	Applications of Fundamental Identities	117–125
6.2	Verifying Trigonometric Identities	114, 117–122, 124
6.3	Solving Trigonometric Equations	109, 110, 112
6.4	Sum and Difference Formulas	117–120
6.5	Multiple-Angle Formulas and Product-Sum Formulas	117–124
Chapter 7	**Additional Applications of Trigonometry**	109–125
7.1	Law of Sines	
7.2	Law of Cosines	
7.3	Vectors	
7.4	Trigonometric Form of Complex Numbers	
7.5	DeMoivre's Theorem and nth Roots	
Chapter 8	**Systems of Equations and Inequalities**	
8.1	Systems of Equations	79–82
8.2	Systems of Linear Equations in Two Variables	79–82
8.3	Systems of Linear Equations in More than Two Variables	79–82
8.4	Systems of Inequalities	45–50, 79–82
8.5	Linear Programming	
Chapter 9	**Matrices and Determinants**	
9.1	Matrices and Systems of Linear Equations	
9.2	Operations with Matrices	
9.3	The Inverse of a Matrix	
9.4	The Determinant of a Matrix	
9.5	Properties of Determinants	
9.6	Applications of Determinants and Matrices	

	Precalculus, 2nd Edition (Section Numbers)	The Algebra of Calculus (Page Numbers)
Chapter 10	**Sequences, Series, and Probability**	
10.1	Sequences and Summation Notation	127–131
10.2	Arithmetic Sequences	127–131
10.3	Geometric Sequences and Series	127, 130
10.4	Mathematical Induction	127–131
10.5	The Binomial Theorem	127
10.6	Counting Principles, Permutations, and Combinations	
10.7	Probability	
Chapter 11	**Some Topics in Analytic Geometry**	
11.1	Introduction to Conics: Parabolas	
11.2	Ellipses	30, 42
11.3	Hyperbolas	
11.4	Rotation and the General Second-Degree Equation	
11.5	Polar Coordinates	117, 124, 125
11.6	Graphs of Polar Equations	109, 124, 125
11.7	Polar Equations of Conics	117, 124, 125
11.8	Plane Curves and Parametric Equations	109, 110, 117, 121

Cross Reference Chart for
Brief Calculus with Applications, 3E
Larson/Hostetler/Edwards, 1991

Brief Calculus with Applications, 3rd Edition (Section Numbers)	The Algebra of Calculus (Page Numbers)
Chapter 0 A Precalculus Review	
0.1 The Real Line and Order	45–52
0.2 Absolute Value and Distance on the Real Line	51–55
0.3 Exponents and Radicals	8–12, 56–57, 74, 88–95
0.4 Factoring Polynomials	8–12
0.5 Fractions and Rationalization	1–12, 90
Chapter 1 Functions, Graphs, and Limits	
1.1 The Cartesian Plane and the Distance Formula	30–34
1.2 Graphs of Equations	30–44
1.3 Lines in the Plane and Slope	25–29, 30–37
1.4 Functions	56–63, 67–70
1.5 Limits	88
1.6 Continuity	56–64
Chapter 2 Differentiation	
2.1 The Derivative and the Slope of a Curve	25–29
2.2 Some Rules for Differentiation	25–29
2.3 Rates of Change: Velocity and Marginals	8–13, 25–29
2.4 The Product and Quotient Rules	1, 8–12
2.5 The Chain Rule	6, 88–92
2.6 Higher-Order Derivatives	89
2.7 Implicit Differentiation	1, 8–12, 25–29
2.8 Related Rates	79, 81
Chapter 3 Applications of the Derivative	
3.1 Increasing and Decreasing Functions	88, 93–94
3.2 Extrema and the First-Derivative Test	8–13, 15–23
3.3 Concavity and the Second-Derivative Test	8–13
3.4 Optimization Problems	13, 18–23, 96–99
3.5 Business and Economics Applications	18–23
3.6 Asymptotes	8–12
3.7 Curve Sketching: A Summary	8–12
3.8 Differentials and Marginal Analysis	13–23, 90–91

	Brief Calculus with Applications, 3rd Edition (Section Numbers)	The Algebra of Calculus (Page Numbers)
Chapter 4	**Integration and Its Applications**	
4.1	Antiderivatives and Indefinite Integrals	88–92
4.2	The General Power Rule	88–92
4.3	Area and the Fundamental Theorem of Calculus	90–91, 96–97
4.4	The Area of a Region Bounded by Two Curves	8–23, 30–43, 67–69, 80–82
4.5	The Definite Integral as the Limit of a Sum	60–62
4.6	Volumes of Solids of Revolution	30–42, 96–97, 100
Chapter 5	**Exponential and Logarithmic Functions**	
5.1	Exponential Functions	88, 93–94
5.2	Exponential Functions: Differentiation and Integration	88
5.3	The Natural Logarithmic Function	104–107
5.4	Logarithmic Functions: Differentiation and Integration	104–107
5.5	Exponential Growth and Decay	94, 104
Chapter 6	**Techniques of Integration**	
6.1	Integration by Substitution	2, 88–91
6.2	Integration by Parts and Present Value	104
6.3	Partial Fractions	3, 79–81
6.4	Integration by Tables and Completing the Square	84–92
6.5	Numerical Integration	45–46, 51–55, 88, 96, 127
6.6	Improper Integrals	4–6, 30, 46, 90–91
Chapter 7	**Probability and Calculus**	
7.1	Discrete Probability	45–56
7.2	Continuous Random Variables	45–46, 56, 63
7.3	Variance and Special Probability Density	46, 96
Chapter 8	**Differential Equations**	
8.1	Solutions and Differential Equations	88
8.2	Separation of Variables	13–24
8.3	First-Order Linear Differential Equations	13–24, 88
8.4	Applications of Differential Equations	13–24, 88

	Brief Calculus with Applications, 3rd Edition (Section Numbers)	The Algebra of Calculus (Page Numbers)
Chapter 9	**Functions of Several Variables**	
9.1	The Three-Dimensional Coordinate System	84–87
9.2	Surfaces in Space	30, 60–62
9.3	Functions of Several Variables	60–62
9.4	Partial Derivatives	13–24, 60–62
9.5	Extrema of Functions of Two Variables	60–62, 79–83
9.6	Lagrange Multipliers and Constrained Optimization	60–62, 79–83
9.7	The Method of Least Squares	60–62, 79–83, 127–132
9.8	Double Integrals and Area in the Plane	3–7, 60–62, 88–95
9.9	Applications of Double Integrals	60–62, 96–103
Chapter 10	**Series and Taylor Polynomials**	
10.1	Sequences	48, 127–131
10.2	Series and Convergence	127–131
10.3	p-Series and the Ratio Test	51–55, 88, 127–131
10.4	Power Series and Taylor's Theorem	127–131
10.5	Taylor Polynomials	127
10.6	Newton's Method	60–62
Chapter 11	**The Trigonometric Functions**	
11.1	Radian Measure of Angles	109–110
11.2	The Trigonometric Functions	109–126
11.3	Graphs of Trigonometric Functions	110
11.4	Derivatives of Trigonometric Functions	109–111, 113–114, 117
11.5	Integrals of Trigonometric Functions	109–111, 113–114, 117

1

Selected Elementary Algebra

REVIEW OF FUNDAMENTALS

■ **Expand** the following expressions in one step.

Expression	Expanded Expression
$2(a + b)$	$2a + 2b$
$\sqrt{a + b}$	not possible
$(a + b)^2$	$a^2 + 2ab + b^2$
$(a - b)^2$	$a^2 - 2ab + b^2$
$(a + b)^3$	$a^3 + 3a^2b + 3ab^2 + b^3$
$(a - b)^3$	$a^3 - 3a^2b + 3ab^2 - b^3$

■ We simplify the **compound fraction**

$$\frac{a/b}{c/d}$$

as follows.

$$\begin{aligned}
\frac{(a/b)}{(c/d)} &= \frac{a}{b} \div \frac{c}{d} \\
&= \frac{a}{b} \cdot \frac{d}{c} \qquad \text{Invert and multiply} \\
&= \frac{ad}{bc}
\end{aligned}$$

■ Express the following as **sums of simpler fractions.**

Expression	Simplified Fractions
$\dfrac{k}{l + m}$	not possible
$\dfrac{n + p}{q}$	$\dfrac{n}{q} + \dfrac{p}{q}$

■ We express

$$\frac{p}{q} + \frac{r}{s}$$

as a **single fraction** using a **least common denominator** (LCD) as follows.

$$\begin{aligned}
\frac{p}{q} + \frac{r}{s} &= \frac{ps}{qs} + \frac{qr}{qs} \qquad \left(\text{LCD of } \frac{p}{q} \text{ and } \frac{r}{s} \text{ is } qs. \right) \\
&= \frac{ps + qr}{qs}
\end{aligned}$$

EXAMPLE 1   **Expanding Powers of Binomials**

Expand the following powers of binomials.

(a) $(5x + 2)^3$ **(b)** $(x^4 - 1)^2$

SOLUTION

(a) Using the formula $(a + b)^3 = a^3 + 3a^2b + 3ab^2 + b^3$ with $a = 5x$ and $b = 2$, we obtain

$$(5x + 2)^3 = (5x)^3 + 3(5x)^2(2) + 3(5x)2^2 + 2^3$$
$$= 5^3x^3 + 3 \cdot 5^2x^2 \cdot 2 + 3 \cdot 5x \cdot 4 + 8$$
$$= 125x^3 + 150x^2 + 60x + 8.$$

(b) Using the formula $(a - b)^2 = a^2 - 2ab + b^2$ with $a = x^4$ and $b = 1$, we obtain

$$(x^4 - 1)^2 = (x^4)^2 - 2 \cdot x^4 \cdot 1 + 1^2$$
$$= x^8 - 2x^4 + 1.$$

SIMILAR PROBLEMS

Expand the following powers of binomials.

1. $(3x + 1)^3$ **2.** $(x^3 - 2)^2$

ANSWERS

1. $27x^3 + 27x^2 + 9x + 1$ **2.** $x^6 - 4x^3 + 4$

EXAMPLE 2   **Substitution**

Substitute $x + h$ for x in the expression $f(x) = x^3 - 7x^2 + x + 1$.

SOLUTION

In the expression $f(x) = x^3 - 7x^2 + x + 1$ we replace x by $x + h$ and obtain the following.

$$f(x + h) = (x + h)^3 - 7(x + h)^2 + (x + h) + 1$$
$$= x^3 + 3x^2h + 3xh^2 + h^3 - 7(x^2 + 2xh + h^2) + x + h + 1$$
$$= x^3 + 3x^2h + 3xh^2 + h^3 - 7x^2 - 14xh - 7h^2 + x + h + 1$$

SIMILAR PROBLEM

1. Substitute $x + h$ for x in the expression $2x^3 + 3x^2 - x + 4$.

ANSWER

1. $2x^3 + 6x^2h + 6xh^2 + 2h^3 + 3x^2 + 6xh + 3h^2 - x - h + 4$

EXAMPLE 3

Common Errors with Fractions

Are the following statements true or false?

(a) $\dfrac{2k}{2x+h} = \dfrac{k}{x+h}$ **(b)** $\dfrac{1}{p+q} = \dfrac{1}{p} + \dfrac{1}{q}$

(c) $\dfrac{x+y}{2} = \dfrac{x}{2} + \dfrac{y}{2}$ **(d)** $3 \cdot \dfrac{a}{b} = \dfrac{a}{3b}$

(e) $3 \cdot \dfrac{a}{b} = \dfrac{3a}{b}$ **(f)** $3 \cdot \dfrac{a}{b} = \dfrac{3a}{3b}$

(g) $3 \cdot \dfrac{a+b}{c} = \dfrac{3a+b}{c}$

SOLUTION

(a) False. In the expression

$$\frac{2k}{2x+h}$$

the 2's cannot be canceled because cancellation is permissible only for factors of the numerator and the denominator. The 2 in the denominator is not a factor of the entire denominator.

(b) False. Try substituting $p = 1$ and $q = 1$. The equation would read

$$\frac{1}{1+1} = \frac{1}{1} + \frac{1}{1} \quad \text{or} \quad \frac{1}{2} = 2 \, !$$

(c) True

(d) False

(e) True

(f) False

(g) False. The line separating the denominator and numerator can be thought of as enclosing both of these quantities in parentheses. Thus,

$$3 \cdot \frac{a+b}{c} = \frac{3}{1} \cdot \frac{a+b}{c} = \frac{3(a+b)}{c} = \frac{3a+3b}{c}.$$

SIMILAR PROBLEMS

Are the following statements true or false?

1. $\dfrac{3r}{3s+t} = \dfrac{r}{s+t}$ **2.** $\dfrac{1}{p-q} = \dfrac{1}{p} - \dfrac{1}{q}$

3. $\dfrac{x-y}{5} = \dfrac{x}{5} - \dfrac{y}{5}$ **4.** $4 \cdot \dfrac{a}{b} = \dfrac{a}{4b}$

5. $4 \cdot \dfrac{a}{b} = \dfrac{4a}{b}$ **6.** $4 \cdot \dfrac{a}{b} = \dfrac{4a}{4b}$

7. $4 \cdot \dfrac{a-b}{c} = \dfrac{4a-b}{c}$

ANSWERS

1. False   2. False
3. True   4. False
5. True   6. False
7. False

EXAMPLE 4

Simplifying Compound Fractions

Simplify the following compound fractions.

(a) $\dfrac{x/2}{x/4}$ (b) $\dfrac{\dfrac{h}{x+h}}{h}$

SOLUTION

(a)
$$\frac{x/2}{x/4} = \frac{x}{2} \div \frac{x}{4} \qquad \text{Invert and multiply}$$
$$= \frac{x}{2} \cdot \frac{4}{x} \qquad \text{Cancel}$$
$$= 2$$

(b)
$$\frac{\dfrac{h}{x+h}}{h} = \frac{h}{x+h} \div h$$
$$= \frac{h}{x+h} \div \frac{h}{1} \qquad \text{Write } h \text{ as } \frac{h}{1}$$
$$= \frac{h}{x+h} \cdot \frac{1}{h} \qquad \text{Invert and multiply}$$
$$= \frac{h}{x+h} \cdot \frac{1}{h} = \frac{1}{x+h} \qquad \text{Cancel}$$

SIMILAR PROBLEMS

Simplify the following compound fractions.

1. $\dfrac{z/3}{z/9}$ 2. $\dfrac{\dfrac{h}{xh+1}}{h}$

ANSWERS

1. 3   2. $\dfrac{1}{xh+1}$

EXAMPLE 5

Simplifying More Fractions

Express $\dfrac{1}{x} + 1$ as a single fraction.

SOLUTION

$$\begin{aligned}
\frac{1}{x} + 1 &= \frac{1}{x} + \frac{1}{1} \qquad\qquad \text{LCD is } x\\
&= \frac{1}{x} + \frac{x}{x}\\
&= \frac{1+x}{x}
\end{aligned}$$

EXAMPLE 6

Simplifying More Fractions

Show that $\dfrac{\dfrac{1}{x+h} - \dfrac{1}{x}}{h} = \dfrac{-1}{x(x+h)}$.

SOLUTION

$$\begin{aligned}
\frac{\dfrac{1}{x+h} - \dfrac{1}{x}}{h} &= \frac{\dfrac{x \cdot 1}{x(x+h)} - \dfrac{1(x+h)}{x(x+h)}}{h} \qquad \text{LCD is } x(x+h)\\[2mm]
&= \frac{\dfrac{x - (x+h)}{x(x+h)}}{h}\\[2mm]
&= \frac{\dfrac{x - x - h}{x(x+h)}}{h}\\[2mm]
&= \frac{\dfrac{-h}{x(x+h)}}{h}\\[2mm]
&= \frac{-h}{x(x+h)} \div h\\[2mm]
&= \frac{-h}{x(x+h)} \cdot \frac{1}{h}\\[2mm]
&= \frac{-h}{x(x+h)} \cdot \frac{1}{h} = \frac{-1}{x(x+h)}
\end{aligned}$$

SIMILAR PROBLEMS

1. Express $\dfrac{1}{y} - 1$ as a single fraction.

2. Show that $\dfrac{\dfrac{1}{y-k} - \dfrac{1}{y}}{k} = \dfrac{1}{y(y-k)}$.

ANSWERS

1. $\dfrac{1-y}{y}$

2. $\dfrac{\dfrac{1}{y-k} - \dfrac{1}{y}}{k} = \dfrac{\dfrac{y}{y(y-k)} - \dfrac{y-k}{y(y-k)}}{k}$

$\qquad\qquad\qquad = \dfrac{y-y+k}{y(y-k)} \div k$

$\qquad\qquad\qquad = \dfrac{k}{ky(y-k)}$

$\qquad\qquad\qquad = \dfrac{1}{y(y-k)}$

EXAMPLE 7

Rationalizing the Numerator

Show that $\dfrac{\sqrt{x+h} - \sqrt{x}}{h} = \dfrac{1}{\sqrt{x+h} + \sqrt{x}}$.

SOLUTION

First multiply by $\left(\sqrt{x+h} + \sqrt{x}\right)/\left(\sqrt{x+h} + \sqrt{x}\right)$.

$$\frac{\sqrt{x+h} - \sqrt{x}}{h} \cdot \frac{\sqrt{x+h} + \sqrt{x}}{\sqrt{x+h} + \sqrt{x}}$$

Now using the formula $(a-b) \cdot (a+b) = a^2 - b^2$, we obtain

$$\frac{\sqrt{x+h} - \sqrt{x}}{h} = \frac{\left(\sqrt{x+h} - \sqrt{x}\right) \cdot \left(\sqrt{x+h} + \sqrt{x}\right)}{h\left(\sqrt{x+h} + \sqrt{x}\right)}$$

$$= \frac{\left(\sqrt{x+h}\right)^2 - \left(\sqrt{x}\right)^2}{h\left(\sqrt{x+h} + \sqrt{x}\right)}$$

$$= \frac{(x+h) - x}{h\left(\sqrt{x+h} + \sqrt{x}\right)}$$

$$= \frac{h}{h\left(\sqrt{x+h} + \sqrt{x}\right)} \qquad \text{Cancel } h$$

$$= \frac{1}{\sqrt{x+h} + \sqrt{x}}.$$

SIMILAR PROBLEM

1. Show that $\dfrac{\sqrt{x} - \sqrt{x + h}}{h} = \dfrac{-1}{\sqrt{x} + \sqrt{x + h}}$.

ANSWER

1. $\dfrac{\sqrt{x} - \sqrt{x + h}}{h} = \dfrac{\sqrt{x} - \sqrt{x + h}}{h} \cdot \dfrac{\sqrt{x} + \sqrt{x + h}}{\sqrt{x} + \sqrt{x + h}}$

$$= \frac{x - (x + h)}{h(\sqrt{x} + \sqrt{x + h})}$$

$$= \frac{-h}{h(\sqrt{x} + \sqrt{x + h})}$$

$$= \frac{-1}{\sqrt{x} + \sqrt{x + h}}$$

CHAPTER 1 EXERCISES

1. Expand $(x + h)^3$.

2. Expand $(x^2 - 3)^3$.

3. Simplify $\dfrac{3x + 2y}{3x + y}$, if possible.

4. Simplify $\dfrac{3x + 2y}{3y}$, if possible.

5. Simplify $\dfrac{5(x + h)^3 - 5x^3}{h}$.

6. Convert $\dfrac{3a}{b} + b$ to a single fraction.

7. Simplify $\dfrac{\dfrac{a}{2x + h} - \dfrac{a}{2x}}{h}$.

8. Simplify $\dfrac{(a/b) - a}{a + (a/b)^2}$.

9. Show that $\dfrac{x + 4}{3x^2 - 7x} = \dfrac{\dfrac{1}{x} + \dfrac{4}{x^2}}{3 - (7/x)}$.

10. Simplify $\dfrac{(x - 1)^2(3x - 1) - 2(x - 1) \cdot 3}{(x - 1)^4}$.

11. Simplify $\dfrac{(x - 1)(2x + 1) - (x + 3)}{(x - 2)(3x + 1)}$.

12. Simplify $\dfrac{[1/(x + h)^2] - [1/x^2]}{h}$.

13. Show that $\dfrac{\sqrt{x + 9} - 3}{x} = \dfrac{1}{\sqrt{x + 9} + 3}$.

14. Show that $\dfrac{\sqrt{4(x + h) - 3} - \sqrt{4x - 3}}{h} = \dfrac{4}{\sqrt{4(x + h) - 3} + \sqrt{4x - 3}}$.

15. Show that $\dfrac{\dfrac{1}{\sqrt{x}} - \dfrac{1}{\sqrt{x + h}}}{h} = \dfrac{\sqrt{x}\sqrt{x + h}}{x(x + h)(\sqrt{x + h} + \sqrt{x})}$.

2 Factoring

REVIEW OF FUNDAMENTALS

■ To **factor** an expression, do the following.

1. Factor out the highest factor common to all of the terms.

2. Then, in most cases, apply one of the following techniques.

Standard Form	Name	Factored Form
$a^2 - b^2$	Difference of Squares	$(a - b)(a + b)$
$a^3 - b^3$	Difference of Cubes	$(a - b)(a^2 + ab + b^2)$
$a^3 + b^3$	Sum of Cubes	$(a + b)(a^2 - ab + b^2)$
$ax^2 + bx + c$	Trinomial	(See Example 2.)

EXAMPLE 1

Factoring

Factor the expression $x^3 + 1$.

SOLUTION

1. Factors common to x^3 and 1 (except 1): none.

2. Type of factoring required: sum of cubes.

$$x^3 + 1 = x^3 + 1^3$$

Now we use the formula

$$a^3 + b^3 = (a + b)(a^2 - ab + b^2)$$

to obtain

$$x^3 + 1^3 = (x + 1)(x^2 - x \cdot 1 + 1^2) \qquad \text{Replace } a \text{ with } x \text{ and } b \text{ with } 1$$
$$= (x + 1)(x^2 - x + 1).$$

SIMILAR PROBLEM

1. Factor $g^3 - 1$.

ANSWER

1. $(g - 1)(g^2 + g + 1)$

EXAMPLE 2

Factoring a Trinomial

Factor the following trinomial.

$$4x^2 - 21x - 18$$

SOLUTION

1. Factors common to all of $4x^2$, $21x$, and 18 (except 1): none.
2. Factoring technique required: trinomial method. We begin by setting up two pairs of parentheses.

$$(\quad)(\quad)$$

For the two first terms choose expressions whose product is the first term of the given trinomial, $4x^2$.

$$(4x\quad)(x\quad)$$

The other choice, $2x$ and $2x$, will be tried if the choice $4x$ and x does not work. Find, by trial and error, the correct two factors of the last term of the given trinomial, -18. In other words, the pair of resulting parentheses will expand to give the original trinomial. For example, consider the following.

$(4x - 1)(x + 18)$ does not work because it multiplies out to $4x^2 + 71x - 18$.

$(4x - 2)(x + 9)$ does not work because it multiplies out to $4x^2 + 34x - 18$.

$(4x + 3)(x - 6)$ does work because it multiplies out to $4x^2 - 21x - 18$.

If none of the possibilities work, try a different factoring for the first term of the trinomial, $4x^2$.

EXAMPLE 3

Factoring a Trinomial

Factor the following trinomial.

$$2x^2 + x - 3$$

SOLUTION

1. Factors common to $2x^2$, x, and 3 (except 1): none.
2. Factoring technique required: trinomial method. Set up the parentheses:

$$(2x\quad)(x\quad)$$

Possible factors of -3 that could fill in the blanks are: 1 and -3, -1 and 3, -3 and 1, and 3 and -1. Only 3 and -1 are correct because when we expand $(2x + 3)(x - 1)$, we do get $2x^2 + x - 3$.

SIMILAR PROBLEMS ■■■■■■■■

Factor the following.

1. $3x^2 + x - 2$ **2.** $4x^2 - 16x + 15$

ANSWERS

1. $(3x - 2)(x + 1)$ **2.** $(2x - 5)(2x - 3)$

EXAMPLE 4
■■■

Factoring a Polynomial

Factor the polynomial $3x^2 + 6x^3 - 9x$.

SOLUTION

1. Highest factor common to $3x^2$, $6x^3$, and $9x$: $3x$. Factor $3x$ out to obtain
$3x^2 + 6x^3 - 9x = 3x(x + 2x^2 - 3)$.

2. Factoring technique required: trinomial method. We choose this method because the expression in parentheses can be rearranged to obtain $3x(x + 2x^2 - 3) = 3x(2x^2 + x - 3)$. Using the trinomial method, we obtain $3x(2x + 3)(x - 1)$. Example 3 gives this last step in greater detail. ■■

SIMILAR PROBLEM ■■■■■■■■

1. Factor $6x^3 + 2x^2 - 4x$.

ANSWER

1. $2x(3x - 2)(x + 1)$

EXAMPLE 5
■■■

Factoring

Factor $(x + 1)^3(4x - 9) - (16x + 9)(x + 1)^2$.

SOLUTION

1. Highest factor common to the terms $(x + 1)^3(4x - 9)$ and $(16x + 9)(x + 1)^2$: $(x + 1)^2$. If there were no common factor, we would expand the expression.

$$(x + 1)^3(4x - 9) - (16x + 9)(x + 1)^2 = (x + 1)^2[(x + 1)(4x - 9) - (16x + 9)]$$
$$= (x + 1)^2[4x^2 - 5x - 9 - 16x - 9]$$
$$= (x + 1)^2[4x^2 - 21x - 18]$$

2. Applying the trinomial method to $4x^2 - 21x - 18$ (see Example 2), we obtain $4x^2 - 21x - 18 = (4x + 3)(x - 6)$. Therefore,

$$(x + 1)^3(4x - 9) - (16x + 9)(x + 1)^2 = (x + 1)^2(4x + 3)(x - 6).$$

■■

SIMILAR PROBLEM ▬▬▬▬▬▬▬▬▬▬▬▬▬▬

1. Factor $(x - 1)^3(2x - 3) - (2x + 12)(x - 1)^2$.

ANSWER

1. $(x - 1)^2(x + 1)(2x - 9)$

EXAMPLE 6 ▬▬▬▬▬▬▬▬   **Simplifying a Quotient**

Simplify the following quotient.

$$\frac{x(5x + 1) - 3(x^2 + 1)}{(x - 1)^2}$$

SOLUTION

To simplify a fraction, first factor the entire numerator and denominator, and then cancel as much as possible. The parentheses in the numerator are of no value in factoring, so we expand them.

$$\frac{x(5x + 1) - 3(x^2 + 1)}{(x - 1)^2} = \frac{5x^2 + x - 3x^2 - 3}{(x - 1)^2} = \frac{2x^2 + x - 3}{(x - 1)^2}$$

In this form, the numerator can now be factored. (See Example 3.)

$$\frac{2x^2 + x - 3}{(x - 1)^2} = \frac{(2x + 3)(x - 1)}{(x - 1)^2}$$

$$= \frac{(2x + 3)(x - 1)}{(x - 1)^2} \qquad \text{Cancel}$$

$$= \frac{2x + 3}{x - 1}$$

EXAMPLE 7 ▬▬▬▬▬▬▬▬   **Simplifying a Quotient**

Simplify the following quotient.

$$\frac{(x + 1)^3(4x - 9) - (16x + 9)(x + 1)^2}{(x - 6)(x + 1)^3}$$

SOLUTION

The parentheses are useful in factoring because the factor $(x+1)^2$ is common to both terms. (See Example 5.) Thus,

$$\frac{(x+1)^3(4x-9)-(16x+9)(x+1)^2}{(x-6)(x+1)^3} = \frac{(x+1)^2[(x+1)(4x-9)-(16x+9)]}{(x-6)(x+1)^3}$$

$$= \frac{(x+1)^2[4x^2-5x-9-16x-9]}{(x-6)(x+1)^3}$$

$$= \frac{[4x^2-21x-18]}{(x-6)(x+1)}$$

$$= \frac{(4x+3)(x-6)}{(x-6)(x+1)} \quad \text{(See Example 2.)}$$

$$= \frac{4x+3}{x+1}$$

SIMILAR PROBLEMS

Simplify the following.

1. $\dfrac{3x(x+1)-2(2x+1)}{(x-1)^2}$

2. $\dfrac{(x-1)^3(2x-3)-(4x-1)(x-1)^2}{(x-1)^2(2x-1)}$

ANSWERS

1. $\dfrac{3x+2}{x-1}$

2. $x-4$

CHAPTER 2 EXERCISES

In Exercises 1–8, factor the expression.

1. y^3-27

2. x^4-16

3. $80x^2+12x-54$

4. $3x^4+5x^3-12x^2$

5. $4x^4-19x^3-5x^2$

6. x^4-x^2-6

7. $4x^4+3x^2-1$

8. $(2x-1)^2(x-3)+(x+1)(2x-1)^3$

In Exercises 9–12, simplify the expression.

9. $\dfrac{x(4-3x)+4(x^2+1)}{x^2-4}$

10. $\dfrac{3x(x-1)-2x^2}{2x^2-5x-3}$

11. $\dfrac{2x(x+1)^2-3(x+1)^3}{8x^3+30x^2+18x}$

12. Factor $x-a$ in such a way that $\sqrt{x}-\sqrt{a}$ is a factor.

3 Equations

REVIEW OF FUNDAMENTALS

■ Linear Equation

Standard Form $px + q = 0$

Example   Solve for y' : $xy' + y = 1 + y'$.   (See Example 1.)

To Identify
- Highest power of the unknown is 1.
- Has no denominators containing the unknown.

To Solve
1. Collect all terms containing the unknown on one side.
2. Factor out the unknown.
3. Divide both sides by the coefficient of the unknown.

■ Quadratic Equation

Standard Form $ax^2 + bx + c = 0$

Example   Solve for x: $2x^2 + x = 3$.   (See Example 5.)

To Identify
- Highest power of the unknown is 2.
- Has no denominators containing the unknown.

More Examples   Solve for x: $x^2 + x + 1 = 0$.   (See Example 4.)
Solve for x: $x^2 - x - 5 = 0$.   (See Example 4.)

To Solve
1. Rewrite all terms on left side in standard form.
2. If left side can be factored, equate factors to zero and solve.
3. If left side is difficult to factor, use the Quadratic Formula
$$x = \frac{-b \pm \sqrt{b^2 - 4ac}}{2a}.$$

■ Polynomial Equation

Standard Form $a_n x^n + a_{n-1} x^{n-1} + ... + a_1 x + a_0 = 0$

Example   Solve for x: $7x^4 - 42x^2 - 35x = 0$.   (See Example 6.)

To Identify
- Highest power of the unknown is 3 or more.
- Has no denominators containing the unknown.

To Solve
1. Rewrite in standard form.
2. If possible, factor left side, equate factors to zero, and solve.

■ General Equation

Examples   Solve for x : $\dfrac{7x^2 + 5x}{x^2 + 1} - \dfrac{5x}{x^2 - 6} = 0$.   (See Example 9.)

Solve for x : $(x + 1)^{-1/2} - \dfrac{2(x - 1)(x + 1)^{1/2}}{\sqrt{4 + 10x + 21x^3}} = 0$.

To Identify
- The equation is neither linear, quadratic, nor polynomial.

To Solve
1. Convert exponents to radicals (used most often).
2. Multiply both sides by the LCD.
3. If a radical remains, isolate it on one side and raise both sides to a power.

EXAMPLE 1

Solving a Linear Equation

Solve the following equation for y'.

$$xy' + y = 1 + y'$$

SOLUTION

This is a linear equation because the unknown y' occurs with power 1 only (and does not occur in the denominator). Since the equation is linear, collect all terms containing the unknown on one side of the equals sign, and collect the remaining terms on the other.

$$xy' + y = 1 + y'$$
$$-y' + xy' = 1 - y$$

Factor out the unknown.

$$y'(-1 + x) = 1 - y$$

Divide both sides of the equation by the coefficient of the unknown. In this example, the coefficient of the unknown is the entire expression $-1 + x$. Dividing by it, we get

$$y' = \frac{1 - y}{-1 + x} \qquad -1 + x \neq 0.$$
$$= \frac{1 - y}{x - 1}.$$

EXAMPLE 2

Solving a Linear Equation

Solve the following equation for y'.

$$2xy^3 + 3x^2y'y^2 + 4 = 3x^2y + x^3yy' + 5y'$$

SOLUTION

The equation is linear in y' because y' occurs with power 1 only and does not occur in a denominator. Since the equation is linear, collect all terms containing the unknown on one side of the equals sign and collect the remaining terms on the other.

$$2xy^3 + 3x^2y'y^2 + 4 = 3x^2y + x^3yy' + 5y'$$
$$-x^3yy' - 5y' + 3x^2y'y^2 = 3x^2y - 2xy^3 - 4$$

Factor out the unknown.

$$y'(-x^3y - 5 + 3x^2y^2) = 3x^2y - 2xy^3 - 4$$

Divide both sides by the coefficient of the unknown.

$$y' = \frac{3x^2y - 2xy^3 - 4}{-x^3y - 5 + 3x^2y^2} \qquad -x^3y - 5 + 3x^2y^2 \neq 0$$
$$= \frac{3x^2y - 2xy^3 - 4}{3x^2y^2 - x^3y - 5}$$

SIMILAR PROBLEMS ■■■■■■■■■■■■■■

Solve for y'.

1. $xy' - y = 1 - y'$ **2.** $3x^2y^2 + x^3 2yy' - 3 = y + xy' + 2y'$

ANSWERS

1. $y' = \dfrac{y+1}{x+1}$ **2.** $y' = \dfrac{y+3-3x^2y^2}{2x^3y-x-2}$

EXAMPLE 3
■■■■■■■■■

Solving a Quadratic Equation

Solve the quadratic equation.

$$4x^2 - 21x - 18 = 0$$

SOLUTION

This is a quadratic equation in standard form. Factor the left side using the trinomial method.

$$4x^2 - 21x - 18 = 0$$
$$(4x+3)(x-6) = 0 \quad \text{Factor left side (See Example 2, page 9.)}$$

Set the factors equal to zero and solve the resulting two equations.

$$\begin{array}{ll} 4x + 3 = 0 & x - 6 = 0 \\ 4x = -3 & x = 6 \\ x = -\dfrac{3}{4} & \end{array}$$

The solutions are $-\frac{3}{4}$ and 6.

■■■■■

SIMILAR PROBLEM ■■■■■■■■■■■■■■

1. Solve for x: $3x^2 + x - 2 = 0$.

ANSWER

1. $\dfrac{2}{3}$, -1

EXAMPLE 4

Solving Quadratic Equations

Solve the following quadratic equations.

(a) $x^2 - x - 5 = 0$ **(b)** $x^2 + x + 1 = 0$

SOLUTION

(a) This is a quadratic equation whose standard form is as follows.

$$ax^2 + bx + c = 0$$
$$1x^2 + (-1)x + (-5) = 0$$

We cannot factor this equation so we use the Quadratic Formula.

$$x = \frac{-b \pm \sqrt{b^2 - 4ac}}{2a}$$

In this example, we get

$$x = \frac{-(-1) \pm \sqrt{(-1)^2 - 4(1)(-5)}}{2(1)}$$
$$= \frac{1 \pm \sqrt{1 + 20}}{2}$$
$$= \frac{1 \pm \sqrt{21}}{2}.$$

This gives two solutions, $(1 + \sqrt{21})/2$ and $(1 - \sqrt{21})/2$.

(b) This is a quadratic equation whose standard form is as follows.

$$ax^2 + bx + c = 0$$
$$(1)x^2 + (1)x + (1) = 0$$

The equation cannot be factored, so we use the Quadratic Formula as follows.

$$x = \frac{-b \pm \sqrt{b^2 - 4ac}}{2a}$$
$$= \frac{-1 \pm \sqrt{1^2 - 4 \cdot 1 \cdot 1}}{2 \cdot 1}$$
$$= \frac{-1 \pm \sqrt{1 - 4}}{2}$$
$$= \frac{-1 \pm \sqrt{-3}}{2}$$

Since the quantity under the square root sign is negative, there is no real solution.

SIMILAR PROBLEMS ▰▰▰▰▰▰▰▰▰▰▰▰▰▰▰▰

Solve for x.

1. $x^2 + 3x - 5 = 0$ **2.** $x^2 - x + 2 = 0$

ANSWERS

1. $\dfrac{-3 + \sqrt{29}}{2}, \ \dfrac{-3 - \sqrt{29}}{2}$ **2.** No real solution

EXAMPLE 5

Solving a Quadratic Equation

Solve the equation $2x - 2 = 1 + x - 2x^2$.

SOLUTION

This is a quadratic equation because the highest power of the unknown is 2, and the equation has no denominator containing the unknown. Since the equation is quadratic, rewrite it in standard form

$$ax^2 + bx + c = 0$$

by collecting all of the terms on the left side and arranging them in descending powers of x.

$$-1 - x + 2x^2 + 2x - 2 = 0$$
$$2x^2 + x - 3 = 0$$

Factor the left side using the trinomial method.

$$(2x + 3)(x - 1) = 0$$

Set the factors equal to zero and solve the two resulting equations.

$$
\begin{array}{ll}
2x + 3 = 0 & \quad x - 1 = 0 \\
2x = -3 & \quad x = 1 \\
x = \dfrac{-3}{2} &
\end{array}
$$

The solutions are $-\frac{3}{2}$ and 1.

▰▰▰▰▰

SIMILAR PROBLEM ▰▰▰▰▰▰▰▰▰▰▰▰▰▰▰▰

1. Solve for x.

$$5x - 1 = 1 + 4x - 3x^2$$

ANSWER

1. $\dfrac{2}{3}$ and -1

EXAMPLE 6   **Solving a Polynomial Equation**

Solve the equation for x.

$$7x^4 - 42x^2 - 35x = 0$$

SOLUTION

This is a polynomial equation because the highest power of the unknown is 3 or more, and there is no denominator containing the unknown. Since the equation is a polynomial, we collect all terms on the left in descending powers of x and factor.

$$7x(x^3 - 6x - 5) = 0$$

Set the factors equal to zero and solve the resulting equations. For the first factor, we have

$$7x = 0$$
$$x = 0.$$

The other equation to be solved is $x^3 - 6x - 5 = 0$. To solve this equation, we use synthetic division on $x^3 - 6x - 5 = 0$. This method works for many of the polynomials encountered in calculus courses. Choose a factor of the constant term of the given polynomial. In this polynomial, the constant term is -5, so we must choose from the factors of -5: 1, -1, 5, and -5. Usually, it is advisable to choose the lowest factor first, but we shall choose 5 for illustrative purposes. List the coefficients of the given polynomial. This polynomial is $1x^3 + 0x^2 + (-6)x + (-5)$, so we list

$$1 \qquad 0 \qquad -6 \qquad -5.$$

Now use the following procedure.

$$
\begin{array}{r|rrrr}
5 & 1 & 0 & -6 & -5 \\
  &   & 5\!*\!* & 25 & 95 \quad \text{* Subtract bottom row from top.} \\
\hline
  & 1\!* & 5 & 19 & 90 \quad \text{** Multiply 5 by 1*}
\end{array}
$$

If the last number is zero, the factoring has been successful. If the last number is not zero, try another factor of the constant term of the polynomial. In this example, the last number is 90, so we try the factor -1 next.

$$
\begin{array}{r|rrrr}
-1 & 1 & 0 & -6 & -5 \\
   &   & -1 & 1 & 5 \\
\hline
   & 1 & -1 & -5 & 0
\end{array}
$$

Since the last number is zero, -1 is a solution of the equation given in this problem. The bottom row of numbers (except 0) gives the coefficients of the new polynomial to be solved. In our example, the bottom row (ignoring the 0) is

$$1 \qquad -1 \qquad -5$$

so we must solve $1x^2 - 1x - 5 = 0$. (See Example 4.)

$$x = \frac{-b \pm \sqrt{b^2 - 4ac}}{2a}$$

$$= \frac{-(-1) \pm \sqrt{(-1)^2 - 4(1)(-5)}}{2(1)}$$

$$= \frac{1 \pm \sqrt{21}}{2}$$

Collecting all solutions, we obtain 0, -1, $(1 + \sqrt{21})/2$, and $(1 - \sqrt{21})/2$. Note that the original polynomial factors as

$$7x^4 - 42x^2 - 35x = 7x(x + 1)(x^2 - x - 5).$$

SIMILAR PROBLEM

1. Solve for x.

$$3x^4 + 6x^3 - 24x^2 + 15x = 0$$

ANSWER

1. 0, 1, $\dfrac{-3 + \sqrt{29}}{2}$, $\dfrac{-3 - \sqrt{29}}{2}$

EXAMPLE 7

Solving a Polynomial Equation

Solve the following equation for x.

$$(x + 1)^3(4x - 9) - (16x + 9)(x + 1)^2 = 0$$

SOLUTION

This equation is a polynomial equation. Since our objective is to factor the left side, we first inspect the parentheses for any common factors which would allow us to factor without performing expansion first. Factoring out $(x+1)^2$, we obtain

$$(x + 1)^2[(x + 1)(4x - 9) - (16x + 9)] = 0$$
$$(x + 1)^2[4x^2 - 5x - 9 - 16x - 9] = 0$$
$$(x + 1)^2[4x^2 - 21x - 18] = 0$$
$$(x + 1)^2(4x + 3)(x - 6) = 0.$$

Set the factors equal to zero and solve.

$$(x + 1)^2 = 0 \qquad 4x + 3 = 0 \qquad x - 6 = 0$$
$$x + 1 = 0 \qquad 4x = -3 \qquad x = 6$$
$$x = -1 \qquad x = -\frac{3}{4}$$

The solutions are -1, $-\frac{3}{4}$, and 6.

SIMILAR PROBLEM ▬▬▬▬▬▬▬▬▬▬▬▬▬▬▬▬▬

1. Solve for x.

$$(x-1)^3(2x-3) - (4x-1)(x-1)^2 = 0$$

ANSWER

1. 1, $\dfrac{1}{2}$, 4

EXAMPLE 8
▬▬▬▬▬▬

Solving a General Equation

Solve the equation for x.

$$\frac{x-1}{(x+1)(3x-5)\sqrt{x+3}} = 0$$

SOLUTION

Caution: Do not set the denominator equal to zero.

$$\frac{x-1}{(x+1)(3x-5)\sqrt{x+3}} = 0 \qquad \text{Multiply both sides by the denominator}$$
$$x - 1 = 0$$
$$x = 1$$

Note: Numbers obtained by this method must be substituted into the denominators of the original equation. Any number which makes a denominator zero is not a solution. Substituting 1 into the denominator

$$(x+1)(3x-5)\sqrt{x+3}$$

we obtain

$$(1+1)(3-5)\sqrt{1+3}.$$

This is not zero, so 1 is a solution of the original equation.

▬▬▬▬▬

SIMILAR PROBLEM ▬▬▬▬▬▬▬▬▬▬▬▬▬▬▬▬▬

1. Solve for x.

$$\frac{x+2}{(x-1)\sqrt{x+3}(2x-3)} = 0$$

ANSWER

1. $x = -2$

EXAMPLE 9   **Solving a General Equation**

Solve the equation for x.

$$\frac{7x^2 + 5x}{x^2 + 1} - \frac{5x}{x^2 - 6} = 0$$

SOLUTION

Since this equation is neither linear, quadratic, nor polynomial, we use the following technique, which works for many of the equations encountered in calculus courses. Multiply each term by the LCD or by the product of the denominators.

$$\frac{7x^2 + 5x}{x^2 + 1} - \frac{5x}{x^2 - 6} = 0$$

$$(x^2 + 1)(x^2 - 6)\frac{(7x^2 + 5x)}{x^2 + 1} - (x^2 + 1)(x^2 - 6)\frac{5x}{x^2 - 6} = 0 \quad \text{Multiply by LCD}$$

$$(x^2 - 6)(7x^2 + 5x) - (x^2 + 1)5x = 0 \quad \text{Cancel}$$

$$7x^4 + 5x^3 - 42x^2 - 30x - 5x^3 - 5x = 0$$

$$7x^4 - 42x^2 - 35x = 0 \quad \text{(See Example 6.)}$$

$$7x(x^3 - 6x - 5) = 0$$

Thus, the solutions are

$$0, \; -1, \; \frac{1 + \sqrt{21}}{2}, \; \text{ and } \; \frac{1 - \sqrt{21}}{2}.$$

SIMILAR PROBLEM

1. Solve for x.

$$\frac{x^2 - 4}{7x - 16} + \frac{1}{x^2 - 4} = 0$$

ANSWER

1. $0, \; 1, \; \dfrac{-1 + \sqrt{29}}{2}, \; \dfrac{-1 - \sqrt{29}}{2}$

EXAMPLE 10   **Using the General Method to Solve an Equation**

Solve the equation for x.

$$(x + 1)^{-1/2} + \frac{(x - 1)(x + 1)^{1/2}}{\sqrt{x^3 + 1}} = 0$$

SOLUTION

Since the equation is neither linear, quadratic, nor polynomial, we use the general method. Convert negative and fractional exponents to radicals.

$$(x+1)^{-1/2} + \frac{(x-1)(x+1)^{1/2}}{\sqrt{x^3+1}} = 0$$

$$\frac{1}{\sqrt{x+1}} + \frac{(x-1)\sqrt{x+1}}{\sqrt{x^3+1}} = 0$$

Multiply each term by the LCD or the product of the denominators.

$$\frac{\sqrt{x+1}\sqrt{x^3+1}}{\sqrt{x+1}} + \frac{(x-1)\sqrt{x+1}\sqrt{x+1}\sqrt{x^3+1}}{\sqrt{x^3+1}} = 0$$

$$\sqrt{x^3+1} + (x-1)(x+1) = 0 \qquad (\sqrt{x+1} \cdot \sqrt{x+1} = x+1)$$

$$\sqrt{x^3+1} + x^2 - 1 = 0$$

Isolate the radical on one side of the equation.

$$\sqrt{x^3+1} = 1 - x^2$$

$$(\sqrt{x^3+1})^2 = (1-x^2)^2 \qquad \text{Square both sides of the equation.}$$

$$x^3 + 1 = 1 - 2x^2 + x^4$$

Collect all the terms on the left of the equation, in descending powers of x, and factor.

$$x^3 + 1 = 1 - 2x^2 + x^4$$

$$-x^4 + x^3 + 2x^2 = 0$$

$$-x^2(x^2 - x - 2) = 0$$

Setting each factor equal to zero, we obtain

$$-x^2 = 0 \qquad x^2 - x - 2 = 0$$

$$x = 0 \qquad (x-2)(x+1) = 0$$

$$x - 2 = 0 \qquad x + 1 = 0$$

$$x = 2 \qquad x = -1.$$

Substitute each of the numbers obtained into the original equation because at one stage in the solution process, we raised each side of the equation to a power. For $x = 0$, we obtain

$$(x+1)^{-1/2} + \frac{(x-1)(x+1)^{1/2}}{\sqrt{x^3+1}} = (0+1)^{-1/2} + \frac{(0-1)(0+1)^{1/2}}{\sqrt{0^3+1}}$$

$$= 1 + \frac{(-1) \cdot 1}{1}$$

$$= 0.$$

Thus, 0 is a solution. For $x = 2$, we obtain

$$(2+1)^{-1/2} + \frac{(2-1)(2+1)^{1/2}}{\sqrt{2^3+1}} = \frac{1}{\sqrt{3}} + \frac{\sqrt{3}}{3} \neq 0.$$

Thus, 2 is not a solution. For $x = -1$, we obtain

$$(-1+1)^{-1/2} + \frac{(-1-1)(-1+1)^{1/2}}{\sqrt{-1^3+1}} = \frac{1}{\sqrt{0}} + \frac{-2\sqrt{0}}{\sqrt{0}} \neq 0.$$

Thus, -1 is not a solution.

SIMILAR PROBLEM

1. Solve for x.

$$(x+2)^{-1/2} + \frac{(x-2)(x+2)^{1/2}}{\sqrt{16-7x^3}} = 0$$

ANSWER

1. 0

CHAPTER 3 EXERCISES

In Exercises 1–3, solve for x.

1. $x^2 = 4$

2. $x^3 = 8$

3. $x^4 = 16$

4. Solve for p: $hp - 1 = q + kp + 6p$

In Exercises 5–12, solve for x.

5. $6x^2 + 5x - 6 = 0$

6. $3x^2 - 13x + 4 = 0$

7. $x^2 - 4x + 4 = 0$

8. $3x^2 - 7x + 1 = 0$

9. $2x^2 + 3x - 3 = 0$

10. $2x^2 + 3x + 3 = 0$

11. $x^2 - 3x + 4 = 0$

12. $\dfrac{3x + 5}{(x - 1)(x^4 + 7)} = 0$

13. Solve for t: $\dfrac{d}{t + r} = \dfrac{5}{t}$.

14. Solve for y_1: $xy_1 + xy_1 y^2 - 3 = 5y_1 + xy$.

15. Solve for $\dfrac{b}{t}$: $z - 4 - s\dfrac{b}{t} - \dfrac{3b}{t} = r^2 \cdot \dfrac{b}{t} - \dfrac{b}{t} + 9$.

16. Solve for y': $xy' - y'y^2 = \dfrac{x^2 yy'}{2} + 2 + 3x + 2xyy'$.

17. Solve for y_1: $\dfrac{x^2 y_1 + y_1}{x - 1} = xyy_1 + x^2 y_1 + 1$.

In Exercises 18–31, solve for x.

18. $2x^4 + 3x^3 - 2x^2 = 0$

19. $(2x + 1)(x - 1)^2 + (x + 5)(2x + 1)^2 = 0$

20. $(x + 1)(2x + 1)^3 - 3(2x + 1)^2(x + 1) = 0$

21. $x^3 + 2x^2 - x - 2 = 0$

22. $x^4 + 3x^3 - x^2 - 3x = 0$

23. $5 + 2x^4 + 9x = 7x^2 + 9x^3$

24. $\dfrac{x}{x + 1} - \dfrac{1}{x + 3} = \dfrac{1}{5x - 1}$

25. $x^4 + 7x^2 - 8x = 0$

26. $x^4 - 2x^3 - 4x^2 - x = 0$

27. $\dfrac{2x(x - 1)}{x^2 - 4} - \dfrac{3x}{x - 1} = 0$

28. $\dfrac{1}{\sqrt{x - 2}} - \dfrac{2(x - 2)^{1/2}}{x + 1} = 0$

29. $\sqrt{x^2 + 5} + x = 5$

30. $8x^2(x^2 + 3)^{-3} - 2(x^2 + 3)^{-2} = 0$

31. $x^{1/3} \dfrac{2}{3}(x - 3)^{-1/3} + (x - 3)^{2/3} \dfrac{1}{3} x^{-2/3} = 0$

4 Slope

REVIEW OF FUNDAMENTALS

■ The **slope** of a nonvertical line represents the number of units a line rises or falls vertically for each unit of horizontal change from left to right.

■ If you know the **coordinates of two points** on the straight nonvertical line, use the following definition of slope to find the slope of the line.

$$m = \frac{y_2 - y_1}{x_2 - x_1}$$

where (x_1, y_1) is either one of the points, and (x_2, y_2) is the other.

■ To get a **rough estimate of the slope** of a line,

1. First try to use the diagram below.
2. Alternatively, choose two points and compare the vertical distance between them with the horizontal distance between them.

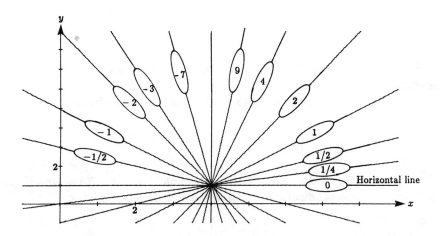

FIGURE 4.1

■ **Slope of perpendicular lines.** If m is the slope of a straight line and $m \neq 0$, then the slope of any perpendicular straight line is $-1/m$. As an example of this, note in Figure 4.1 that the lines with slope 2 and $-1/2$ are perpendicular.

EXAMPLE 1   **Finding the Slope from a Given Pair of Points**

Find the slope of the straight line that contains the two points $(1, -3)$ and $(-2, 5)$.

SOLUTION

In this example, we let $(x_1, y_1) = (1, -3)$, so that $x_1 = 1$ and $y_1 = -3$, which implies that the other point is $(x_2, y_2) = (-2, 5)$. Thus, the slope is

$$\begin{aligned} m &= \frac{y_2 - y_1}{x_2 - x_1} \\ &= \frac{5 - (-3)}{(-2) - 1} \\ &= \frac{8}{-3} \\ &= -\frac{8}{3}. \end{aligned}$$

EXAMPLE 2   **Finding the Slope of a Perpendicular Line**

Find the slope of all straight lines perpendicular to the line described in Example 1.

SOLUTION

The slope of the line in Example 1 is $-\frac{8}{3}$. Thus, the slope of a line that is perpendicular to the line given in Example 1 is

$$\begin{aligned} m &= -\left(\frac{1}{-8/3}\right) \\ &= (-1) \cdot \left(-\frac{3}{8}\right) \\ &= \frac{3}{8}. \end{aligned}$$

SIMILAR PROBLEMS

1. Find the slope of the straight line that contains the two points $(-3, 7)$ and $(-4, -6)$.

2. Find the slope of all straight lines perpendicular to the line described in Problem 1.

ANSWERS

1. 13   **2.** $-\dfrac{1}{13}$

EXAMPLE 3   **Estimating Slope**

Each of the four straight lines shown has a slope that is approximately equal to one of the following values: -50, 0, 0.1, -1, -0.1, $\frac{1}{4}$, 2, 3, and $-\frac{1}{4}$. Choose the correct value for each line.

(a)
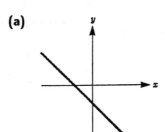

(b)
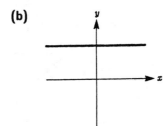

(c)
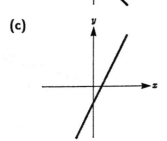

(d)
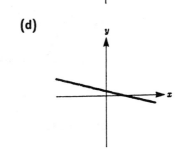

SOLUTION

(a) The slope is clearly negative in this case, so we must choose from -50 (steep), -1 ($45°$), $-\frac{1}{4}$ (shallow), and -0.1 (shallow). We can see that the line makes roughly a $45°$ angle with the x-axis, so the answer is -1.

(b) The line is horizontal, so we get 0.

(c) The line is not horizontal, and its slope is positive, so the possible values are 0.1, $\frac{1}{4}$, 2, and 3. Choose any two convenient points on the line. In doing so, you will see that the vertical distance between the two points is approximately 2 times the horizontal distance between them. Thus, the slope is approximately 2.

(d) The line has a negative shallow slope, so the only values from the given list that might be valid are -0.1 and $-\frac{1}{4}$. Choose any two convenient points on the line. In doing so, you will see that the vertical distance is about $\frac{1}{4}$ times the horizontal distance between the two points and thus, the slope is approximately $-\frac{1}{4}$.

SIMILAR PROBLEMS ▬▬▬▬▬▬▬▬▬▬▬

Estimate the slopes of the lines shown. Choose from among the following values: -5, -1, -0.1, $\frac{1}{4}$, 10, $\frac{1}{2}$, -10, 0, $\frac{1}{10}$, and -2.

1.

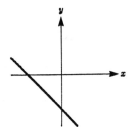

2.

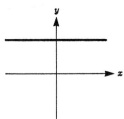

3.

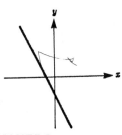

4.

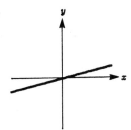

ANSWERS

1. -1

2. 0

3. -2

4. $\frac{1}{4}$

▬▬▬▬▬▬▬▬▬▬▬▬▬▬▬▬▬▬▬▬▬▬

CHAPTER 4 EXERCISES

1. Which of the two straight lines shown has the greater slope?

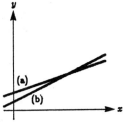

2. Find an expression for the slope of the line that contains the points $(5, q)$ and $(q, -1)$, where $q \neq 5$.

3. If straight line L is perpendicular to straight line T, and the slope of T is $\frac{3}{4}$, what is the slope of L?

4. Estimate the slopes of the given straight lines. Choose from the following values: 2, 0, $\frac{1}{10}$, $\frac{1}{4}$, $-\frac{1}{10}$, -1, 1, and $\frac{1}{2}$.

(a)

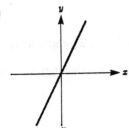

(b)

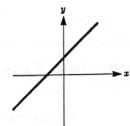

(c)

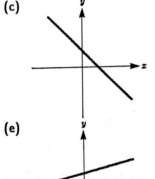

(d)

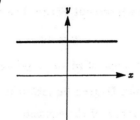

(e)

5. What is the slope of the straight line that intersects the x-axis at the point 3 and the y-axis at the point 2?

5 Common Graphs and Their Equations

REVIEW OF FUNDAMENTALS

■ **Linear Equation**

The general form is $Ax + By + C = 0$. The graph is **a straight line**.

Point-Slope Form: The equation of the line passing through (x_1, y_1) with a slope of m is

$$y - y_1 = m(x - x_1).$$

Slope-Intercept Form: The line given by

$$y = mx + b$$

has a slope of m and a y-intercept at $(0, b)$.

■ **Second-Degree Equation**

The graph of the equation

$$y = ax^2 + bx + c$$

is called a **parabola.** If a is positive, then the parabola opens up. If a is negative, then the parabola opens down.

■ **Other Special Equations and Their Graphs**

$y = \dfrac{1}{x^n}$, n odd

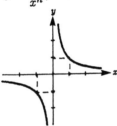

$y = \dfrac{1}{x^n}$, n even

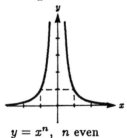

$y = x^n$, n odd

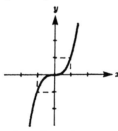

$y = x^n$, n even

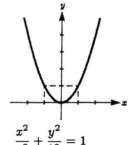

$x^2 + y^2 = r^2$

$\dfrac{x^2}{a^2} + \dfrac{y^2}{b^2} = 1$

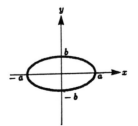

EXAMPLE 1   **The Meaning of "The Graph of an Equation"**

The graph of the equation $y = x^3 - x$ is sketched in Figure 5.1. Answer the following questions precisely (i.e., do not estimate the answers from the graph).

(a) Is the point $(3, 2)$ on the graph?

(b) Is the point $(2, 6)$ on the graph?

(c) What is the distance marked c?

(d) What is the y-coordinate of P?

(e) What are the coordinates of the points at which the curve intersects the x-axis?

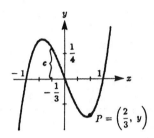

FIGURE 5.1

SOLUTION

(a) To check whether the point $(3, 2)$ lies on the graph, we check whether $x = 3$ and $y = 2$ satisfies the equation.

$$y = x^3 - x$$
$$2 \neq 3^3 - 3$$

Thus, $(3, 2)$ does not lie on the graph.

(b) To check whether the point $(2, 6)$ lies on the graph, we check whether $x = 2$ and $y = 6$ satisfies the equation.

$$y = x^3 - x$$
$$6 = 2^3 - 2$$
$$6 = 6$$

Thus $(2, 6)$ does lie on the graph.

(c) The distance c is the value of y in the equation when x has the value $-\frac{1}{3}$.

$$y = x^3 - x$$
$$= \left(-\frac{1}{3}\right)^3 - \left(-\frac{1}{3}\right)$$
$$= -\frac{1}{27} + \frac{1}{3}$$
$$= \frac{8}{27}$$

(d) The y-coordinate of P is the value of y when $x = \frac{2}{3}$.

$$y = x^3 - x$$

$$= \left(\frac{2}{3}\right)^3 - \left(\frac{2}{3}\right)$$

$$= \frac{8}{27} - \frac{2}{3}$$

$$= -\frac{10}{27}$$

(e) The coordinates of the points at which the curve intersects the x-axis are the values obtained for x when y is set equal to zero.

$$x^3 - x = 0$$

$$x(x^2 - 1) = 0$$

$$x(x - 1)(x + 1) = 0$$

$$x = 0, \ 1, \ -1$$

Thus, the points at which the curve intersects the x-axis are -1, 0, and 1.

SIMILAR PROBLEMS

The graph of the equation $y = x^3 + x^2 - 2x$ is given in Figure 5.2. Answer the following questions precisely (i.e., do not estimate from the graph).

1. Is the point $(1, 1)$ on the graph?

2. Is the point $(1, 0)$ on the graph?

3. What is the distance marked c?

4. What is the y-coordinate of P?

5. What are the coordinates of the points at which the curve intersects the x-axis?

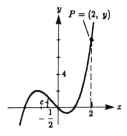

FIGURE 5.2

ANSWERS

1. no   **2.** yes   **3.** $-\frac{5}{8}$
4. 8   **5.** 0, 1, and -2

EXAMPLE 2   **Finding the Equation of a Straight Line**

Find the equation of the straight line that passes through the point $(2, 4)$ and is parallel to the straight line $2x + 3y - 8 = 0$.

SOLUTION

First, we obtain the slope of the line we are seeking from the slope of the line $2x + 3y - 8 = 0$. (Since these two lines are parallel, their slopes are the same.) To find the slope $2x + 3y - 8 = 0$, we rewrite in the slope-intercept form $y = mx + b$. This means that we solve for y.

$$3y = -2x + 8$$
$$y = -\frac{2}{3}x + \frac{8}{3}$$
$$y = mx + b$$

When the equation is written *exactly* in this form, the slope is simply the coefficient of x: $-\frac{2}{3}$. The line we seek also passes through the point $(2, 4)$, and so we can use the point-slope form.

$$y - y_1 = m(x - x_1)$$
$$y - 4 = -\frac{2}{3}(x - 2)$$
$$3y - 12 = -2(x - 2)$$
$$3y = -2x + 4 + 12$$
$$2x + 3y = 16$$

EXAMPLE 3   **Finding the Equation of a Straight Line Perpendicular to a Given Line**

Find the equation of the straight line that is perpendicular to the line $2x + 3y - 8 = 0$ at the point $(1, 2)$.

SOLUTION

The slope of the line $2x + 3y - 8 = 0$ is $m = -\frac{2}{3}$, as shown in Example 2. Thus, the slope of a line perpendicular to this line is $\frac{3}{2}$. The equation of the line passing through $(1, 2)$ is

$$y - 2 = \frac{3}{2}(x - 1)$$
$$4y - 4 = 3(x - 1)$$
$$4y = 3x - 3 + 4$$
$$-3x + 4y = 1$$
$$3x - 4y = -1.$$

SIMILAR PROBLEM

1. Find the equation of the straight line that passes through the point $(1, 3)$ and is parallel to the straight line $3x + 4y - 7 = 0$.

2. Find the equation of the straight line that is perpendicular to the line $3x + 4y - 7 = 0$ at the point $(1, 2)$.

ANSWERS

1. $3x + 4y - 15 = 0$ 2. $4x - 3y + 2 = 0$

EXAMPLE 4

Deriving the Equation of the Straight Line

The straight line with slope 5 that passes through the point $(-1, 3)$ intersects the x-axis at a certain point. What are the coordinates of this point?

SOLUTION

First we find the equation of the straight line. This can be done using the point-slope formula.

$$y - 3 = 5[x - (-1)]$$
$$y - 3 = 5(x + 1)$$
$$y = 5x + 5 + 3$$
$$y = 5x + 8$$

Next, to find the point of intersection with the x-axis, we set $y = 0$ in this equation and solve for x.

$$0 = 5x + 8$$
$$-5x = 8$$
$$x = -\frac{8}{5}$$

EXAMPLE 5

Deriving the Equation of the Straight Line

What are the coordinates of the point at which the straight line passing through the points $(1, -3)$ and $(-2, -4)$ intersects the y-axis?

SOLUTION

First we find the equation of the straight line. The slope of the line passing through $(1, -3)$ and $(-2, -4)$ is

$$m = \frac{-3 - (-4)}{1 - (-2)} = \frac{-3 + 4}{1 + 2} = \frac{1}{3}.$$

Thus, the equation of the line is

$$y - (-3) = \frac{1}{3}(x - 1)$$
$$y + 3 = \frac{1}{3}(x - 1)$$
$$3y + 9 = x - 1$$
$$-x + 3y = -10.$$

Next, to find the point of intersection with the y-axis, we set $x = 0$ in this equation and solve for y.

$$3y = -10$$
$$y = -\frac{10}{3}$$

SIMILAR PROBLEMS

1. The straight line with slope 4 that passes through the point $(2, -1)$ intersects the x-axis at a certain point. What is the coordinate of this point?

2. What is the coordinate of the point at which the straight line passing through the points $(1, -5)$ and $(-1, -3)$ intersects the y-axis?

ANSWERS

1. $\frac{9}{4}$

2. -4

EXAMPLE 6

Sketching a Straight Line Using the 2-Intercept Method

Sketch the graphs of $2x - 3y - 12 = 0$.

SOLUTION

Since the equation is linear, its graph is a straight line. To find its y-intercept, we let $x = 0$ and solve for y.

$$0 - 3y - 12 = 0$$
$$-3y = 12$$
$$y = -\frac{12}{3} = -4$$

To find the x-intercept, we let $y = 0$ and solve for x.

$$2x - 0 - 12 = 0$$
$$2x = 12$$
$$x = \frac{12}{2} = 6$$

Now we plot the two intercepts obtained and join them with a straight line as shown in Figure 5.3.

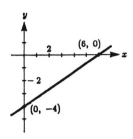

FIGURE 5.3

EXAMPLE 7

Sketching a Straight Line

Sketch the graph of $x = 3$.

SOLUTION

The graph of this equation is a vertical line, as shown in Figure 5.4.

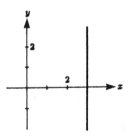

FIGURE 5.4

SIMILAR PROBLEMS

Sketch the graph of the given equation.

1. $3x - 2y + 6 = 0$ **2.** $x = -2$

ANSWERS

1.

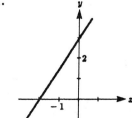

2.

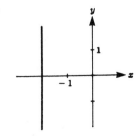

EXAMPLE 8

Sketching a Straight Line Using the Slope-Intercept Method

Sketch the graph of $y = 2x$.

SOLUTION

Using the slope-intercept method compare the equation with the standard form.

$$y = mx + b$$
$$y = 2x + 0$$

Thus, the slope is $m = 2$ and the y-intercept is $(0, 0)$. The graph is shown in Figure 5.5.

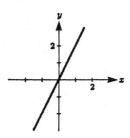

FIGURE 5.5

EXAMPLE 9

Sketching a Straight Line Using the Slope-Intercept Method

Sketch the graph at $y = -x + 3$.

SOLUTION

Using the slope-intercept method, compare the equation with the standard form.

$$y = mx + b$$
$$y = (-1)x + 3$$

Thus, the slope is $m = -1$ and the y-intercept is $(0, 3)$. The graph is shown in Figure 5.6.

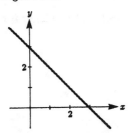

FIGURE 5.6

SIMILAR PROBLEMS ▬▬▬▬▬▬▬▬▬▬▬

Sketch the graphs of the equations.

1. $y = 3x$

2. $y = 2x - 1$

ANSWERS

1.

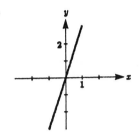

2.

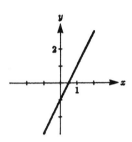

EXAMPLE 10

Sketching a Parabola

Sketch the graph of the equation $y = 2x^2 + x - 3$.

SOLUTION

Because this equation is of the form $y = ax^2 + bx + c$, its graph is a parabola. The parabola opens up because $a = 2$ is positive. To find the y-intercept, we let $x = 0$ and solve for y.

$$y = 2x^2 + x - 3$$
$$y = 0 + 0 - 3$$
$$y = -3$$

Thus, the y-intercept is $(0, -3)$. To find the x-intercepts, we let $y = 0$ and solve for x.

$$2x^2 + x - 3 = 0$$
$$(2x + 3)(x - 1) = 0$$
$$x = -\frac{3}{2}, \ 1$$

Thus, the x-intercepts are $\left(-\frac{3}{2}, 0\right)$ and $(1, 0)$. The graph is shown in Figure 5.7.

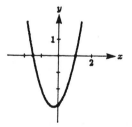

FIGURE 5.7   ▬▬▬▬▬

SIMILAR PROBLEM

1. Sketch the graph of $y = x^2 + x - 2$.

ANSWER

1.

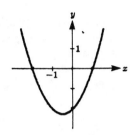

EXAMPLE 11

Sketching a Parabola

Sketch the graph of $y = -x^2 + 4x - 4$.

SOLUTION

Because this equation is of the form $y = ax^2 + bx + c$, its graph is a parabola. The parabola opens down because $a = -1$ is negative. To find the y-intercept, we let $x = 0$ and solve for x.

$$y = -x^2 + 4x - 4$$
$$y = 0 + 0 - 4$$
$$y = -4$$

Thus, the y-intercept is $(0, -4)$. To find the x-intercept(s), we let $y = 0$ and solve for x.

$$-x^2 + 4x - 4 = 0$$
$$x^2 - 4x + 4 = 0$$
$$(x - 2)^2 = 0$$
$$x = 2$$

Thus, there is only one x-intercept and it occurs at $(2, 0)$. The graph is shown in Figure 5.8.

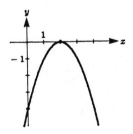

FIGURE 5.8

EXAMPLE 12   **Sketching a Parabola**

Sketch the graph of $y = -\dfrac{x^2}{3}$.

SOLUTION

Because this equation is of the form $y = ax^2 + bx + c$, its graph is a parabola. The parabola opens down because $a = -\frac{1}{3}$ is negative. The only intercept is $(0, 0)$ and the graph is shown in Figure 5.9.

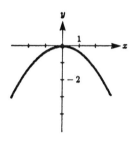

FIGURE 5.9

SIMILAR PROBLEMS

Sketch the following graphs.

1. $y = -x^2 + 6x - 9$ **2.** $y = -5x^2$

ANSWERS

1.

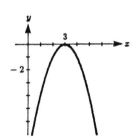

2.

EXAMPLE 13   **Sketching a Parabola**

Sketch the graph of the equation $y = -x^2 - x - 1$.

SOLUTION

Because this equation is of the form $y = ax^2 + bx + c$, its graph is a parabola. The parabola opens down because $a = -1$ is negative. The y-intercept occurs at $(0, -1)$. There are no x-intercepts because the equation $-x^2 - x - 1 = 0$ has no real solutions (See Example 4(b), page 16.). The graph is shown in Figure 5.10.

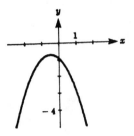

FIGURE 5.10

SIMILAR PROBLEM

1. Sketch $y = x^2 - x + 1$.

ANSWER

1.

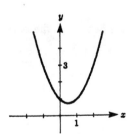

EXAMPLE 14   **Sketching a Circle**

Sketch the graph of $x^2 + y^2 = 3$.

SOLUTION

The equation $x^2 + y^2 = 3$ can be put in the form

$$x^2 + y^2 = r^2$$
$$x^2 + y^2 = (\sqrt{3})^2.$$

Thus, the graph is a circle with radius $\sqrt{3} \approx 1.7$, and center at the origin, as shown in Figure 5.11.

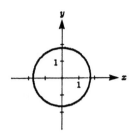

FIGURE 5.11   ▬▬▬

EXAMPLE 15
▬▬▬

Sketching an Ellipse

Sketch the graph of $5x^2 + 2y^2 = 10$.

SOLUTION

The equation $5x^2 + 2y^2 = 10$ can be put in the form

$$\frac{x^2}{a^2} + \frac{y^2}{b^2} = 1$$

$$\frac{x^2}{2} + \frac{y^2}{5} = 1 \qquad \text{Divide by 10}$$

$$\frac{x^2}{(\sqrt{2})^2} + \frac{y^2}{(\sqrt{5})^2} = 1.$$

Thus, the graph is an ellipse with center at the origin, as shown in Figure 5.12.

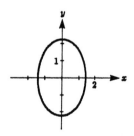

FIGURE 5.12   ▬▬▬

SIMILAR PROBLEMS ▬▬▬▬▬▬▬▬▬

Sketch the graphs

1. $x^2 + y^2 = 2$

2. $3x^2 + 7y^2 = 21$

ANSWERS

1.

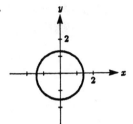

2.

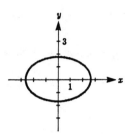

CHAPTER 5 EXERCISES

1. The graph of the equation $y = 2x^3 - 3x^2 - 2x + 3$ is shown below.

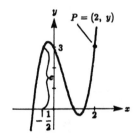

$P = (2, y)$

a. Determine exactly the distance a.
b. Determine exactly the y-coordinate for P.
c. What are the exact coordinates of the points at which the curve intersects the x-axis?

In Exercises 2–10, sketch the graph of the given equation.

2. $2x + 3y = x - 1$

3. $y = -3$

4. $y = x^2 - 6x + 9$

5. $y = 6x^2 - 5x - 6$

6. $y = 2x^2 + x + 1$

7. $y = x^5$

8. $y = \dfrac{1}{x^4}$

9. $x^2 + y^2 = \dfrac{1}{2}$

10. $4y^2 = 36 - 9x^2$

In Exercises 11–15, find the equation of the given line.

11. The straight line with slope -2 passing through the point $(-3, -5)$

12. The straight line passing through $(1, -3)$ that intersects the x-axis at the point whose coordinate is 5

13. The straight line passing through the point $(-1, -2)$ that is parallel to the straight line $x + y = 2x - y + 3$

14. The straight line perpendicular to $x + y = 2x + 3$ that intersects the y-axis at the point whose coordinate is 2

15. Find the equations of the indicated lines.

(a)

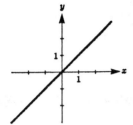

(b)

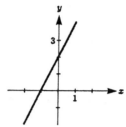

(c)

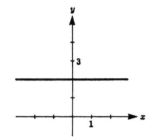

(d)

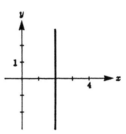

(e)

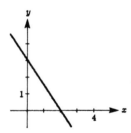

16. Find the equation of the circle shown.

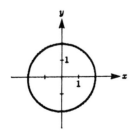

17. Find the equation of the ellipse shown.

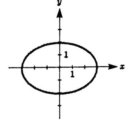

6

Inequalities and Absolute Values

REVIEW OF FUNDAMENTALS

■ **Inequalities:** The inequalities $a < b$ (read a is less than b) means a is to the left of b on the number line. (See Figure 6.1.) The formal definition is $a < b$ *if and only if $b - a$ is positive.* By $a \le b$ we mean that either $a < b$ or $a = b$.

FIGURE 6.1

■ **Operations with Inequalities:** Treat inequalities as you would treat equations when adding to or multiplying both sides, *except* that the inequality sign must be reversed when multiplying or dividing both sides by a negative number.

Operations on the Inequality $a < b$:

Operation	*Result*
Adding to both sides	$a + c < b + c$ $a - c < b - c$
Multiplying both sides	$ac < bc$ if c is positive $ac > bc$ if c is negative
Reciprocating both sides	$\dfrac{1}{a} > \dfrac{1}{b}$ if ab is positive $\dfrac{1}{a} < \dfrac{1}{b}$ if ab is negative

■ **Absolute Value:** The absolute value of a number x, $|x|$ is its distance from 0 along the number line. Precisely, we define $|x|$ as follows.

$$|x| = \begin{cases} x, & \text{if } x \text{ is nonnegative} \\ -x, & \text{if } x \text{ is negative} \end{cases}$$

PROPERTIES OF ABSOLUTE VALUE

	Property
Negation:	$|-a| = |a|$
Products and quotients:	$|ab| = |a|\,|b|$ $\left\lvert\dfrac{a}{b}\right\rvert = \dfrac{|a|}{|b|}$
Triangle Inequality:	$|a + b| \le |a| + |b|$

■ **Set notation:** Set notation is written as follows.

$$\{a : \ldots\} \qquad \text{the set of all } a\text{'s such that} \ldots$$

Unless otherwise specified, a is assumed to be a real number.

■ **Interval notation:** Interval notation denotes the solution set of inequalities.

Set notation	*Interval notation*
$\{x : a < x < b\}$	(a, b)
$\{x : a \leq x \leq b\}$	$[a, b]$
$\{x : x \geq a\}$	$[a, \infty)$

The intervals $[a, b)$, $(a, b]$, (a, ∞), $(-\infty, a]$, and $(-\infty, a)$ are defined similarly.

■ **Relating absolute value and inequality:** The following statements are equivalent.

(a) $|x - 3| < \delta$

(b) $-\delta < x - 3 < \delta$

(c) $3 - \delta < x < 3 + \delta$

(d) x is in the open interval from $3 - \delta$ to $3 + \delta$.

EXAMPLE 1

Understanding Inequalities

Are the following statements true or false?

(a) $3 < 5$ (b) $3 \leq 5$ (c) $3 \leq 3$

(d) $5 < 1$ (e) $1 > 5$ (f) $5 > 1$

(g) $-2 < 5$ (h) $-7 < -2$

SOLUTION

(a) $3 < 5$ is true because 3 is to the left of 5 on the number line.

(b) $3 \leq 5$ is true because $3 < 5$. However, $3 \neq 5$.

(c) $3 \leq 3$ is true because $3 = 3$. However, $3 \not< 3$.

(d) $5 < 1$ is false because 5 is not to the left of 1 on the number line.

(e) $1 > 5$ is false. (This is the same as the statement $5 < 1$ above.)

(f) $5 > 1$ is true because 5 is to the right of 1 on the number line.

(g) $-2 < 5$ is true because -2 is to the left of 5 on the number line.

(h) $-7 < -2$ is true because -7 is to the left of -2 on the number line.

EXAMPLE 2

Understanding Inequalities

Are the following statements true for *all* values of x?

(a) $x^3 + 1 > x^3$ (b) $x^3 + x > x^3$

(c) $2x \geq x$ (d) $x^2 \geq 0$

(e) $x^2 \geq x$ (f) $\sqrt{x} \geq 0$

(g) $-x \leq 0$ (h) $\dfrac{1}{x} \leq x$

SOLUTION

(a) $x^3 + 1 > x^3$ is true because for any value of x^3, adding 1 yields a number that is to the right of x^3 on the number line.

(b) $x^3 + x > x^3$ is false because it is not true for *all* possible values of x. For instance, when $x = -1$, we obtain the following.

$$x^3 + x > x^3$$
$$(-1)^3 + (-1) > (-1)^3$$
$$(-1) + (-1) > -1$$
$$-2 > -1 \qquad \text{False}$$

(c) $2x \geq x$ is false because when $x = -1$, we obtain the following.

$$2x \geq x$$
$$2(-1) \geq -1$$
$$-2 \geq -1 \qquad \text{False}$$

(d) $x^2 \geq 0$ is true because squaring a number yields either zero or a positive number. For example: $3^2 = 9$ is positive; $(-5)^2 = 25$ is positive.

(e) $x^2 \geq x$ is false because when $x = \frac{1}{2}$, we obtain the following.

$$x^2 \geq x$$
$$(\tfrac{1}{2})^2 \geq \tfrac{1}{2}$$
$$\tfrac{1}{4} \geq \tfrac{1}{2} \qquad \text{False}$$

(f) $\sqrt{x} \geq 0$ is true because the expression $\sqrt{x}$ refers specifically to the *nonnegative* number whose square is x. (For example: $\sqrt{9} = 3$, not -3.)

(g) $-x \leq 0$ is false because when $x = -1$, we obtain the following.

$$-x \leq 0$$
$$-(-1) \leq 0$$
$$1 \leq 0 \qquad \text{False}$$

(h) $\dfrac{1}{x} \leq x$ is false because when $x = \frac{1}{2}$, we obtain the following.

$$\frac{1}{x} \leq x$$
$$\frac{1}{1/2} \leq \frac{1}{2}$$
$$2 \leq \frac{1}{2} \qquad \text{False}$$

SIMILAR PROBLEMS

1. Are the following statements true or false?

(a) $2 < 7$ (b) $2 \leq 7$ (c) $4 \leq 4$

(d) $7 < 2$ (e) $2 > 7$ (f) $7 > 1$

(g) $-2 < 7$ (h) $-7 < -4$

2. Are the following statements true for all values of x?

(a) $x^5 + 2 > x^5$ (b) $x^2 + x > x^2$ (c) $3x > x$

(d) $x^4 \geq 0$ (e) $x^4 \geq x$ (f) $\sqrt{x} \geq 0$

(g) $-2x < 0$ (h) $\dfrac{-2}{x} < 0$

ANSWERS

1. (a) true   (b) true   (c) true

(d) false   (e) false   (f) true

(g) true   (h) true

2. (a) true   (b) false   (c) false

(d) true   (e) false   (f) true

(g) false   (h) false

EXAMPLE 3

Solving Linear Inequalities

Solve the following inequalities for x.

(a) $x - 4 < 1$ (b) $3 - 2x \leq 1$ (c) $-1 < x - 4 < 1$

SOLUTION

(a) We treat $x - 4 < 1$ as we would the equation $x - 4 = 1$.

$$x - 4 < 1$$
$$x - 4 + 4 < 1 + 4 \qquad \text{Add 4}$$
$$x < 5$$

(b) We treat $3 - 2x \leq 1$ as we would the equation $3 - 2x = 1$.

$$3 - 2x \leq 1$$
$$3 - 3 - 2x \leq 1 - 3 \qquad \text{Subtract 3}$$
$$-2x \leq -2$$
$$\frac{-2x}{-2} \geq \frac{-2}{-2} \qquad \text{Divide by } -2$$
$$x \geq 1$$

(c) It is often possible to deal with a double inequality as a single entity.

$$-1 < x - 4 < 1$$
$$-1 + 4 < x - 4 + 4 < 1 + 4 \qquad \text{Add 4}$$
$$3 < x < 5$$

SIMILAR PROBLEMS ▬▬▬▬▬▬▬▬▬▬▬▬▬▬▬▬▬▬

Solve for x.

1. $x - 3 < 1$ **2.** $5 - 4x \leq 1$ **3.** $-1 < x - 3 < 1$

ANSWERS

1. $x < 4$ **2.** $x \geq 1$ **3.** $2 < x < 4$

EXAMPLE 4

Solving Quadratic Inequalities

Solve the following inequalities for x.

(a) $2x^2 + x - 3 < 0$ **(b)** $2x^2 + x - 3 > 0$

Quadratic inequalities are inequalities of the form $ax^2 + bx + c > 0$ or < 0 or ≥ 0 or ≤ 0. We can solve them by inspecting the graph of the equation $y = ax^2 + bx + c$.

SOLUTION

(a) To solve $2x^2 + x - 3 < 0$, we first sketch $y = 2x^2 + x - 3$, as shown in Figure 6.2. The graph of $y = 2x^2 + x - 3 = (2x + 3)(x - 1)$ is a parabola with x-intercepts at $\left(-\frac{3}{2}, 0\right)$ and $(1, 0)$. From Figure 6.2, we see that $2x^2 + x - 3 < 0$ when $-\frac{3}{2} < x < 1$.

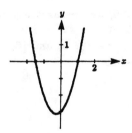

FIGURE 6.2

(Notice that since the value of $2x^2 + x - 3$ equals zero when $x = 1$ and $-\frac{3}{2}$ the answer does not include the number 1 or $-\frac{3}{2}$.)

(b) To solve $2x^2 + x - 3 > 0$, we use the graph of $y = 2x^2 + x - 3$ shown in Figure 6.2 and conclude that the solution consists of two intervals:

$$\left(-\infty, -\frac{3}{2}\right) \quad \text{and} \quad (1, \infty).$$

EXAMPLE 5

Solving a Quadratic Inequality

Solve the inequality $2x^2 + 4x > 3x + 3$.

SOLUTION

We first treat $2x^2 + 4x > 3x + 3$ as we would the equation $2x^2 + 4x = 3x + 3$.

$$2x^2 + 4x > 3x + 3$$
$$2x^2 + 4x - 3x - 3 > 3x + 3 - 3x - 3 \qquad \text{Subtract } 3x + 3$$
$$2x^2 + x - 3 > 0$$

This inequality is the same as that given in part (b) of Example 4. Hence, the solution is

$$\left(-\infty, \ -\frac{3}{2}\right) \bigcup (1, \ \infty).$$

EXAMPLE 6

Solving Quadratic Inequalities

Solve the following inequalities.

(a) $4x^2 - 21x - 18 < 0$ **(b)** $4x^2 - 21x - 18 \le 0$

SOLUTION

(a) To solve $4x^2 - 21x - 18 < 0$, we first sketch $y = 4x^2 - 21x - 18$, as shown in Figure 6.3. The graph of $y = 4x^2 - 21x - 18 = (4x + 3)(x - 6)$ is a parabola with x-intercepts at $\left(-\frac{3}{4}, \ 0\right)$ and $(6, 0)$. From Figure 6.3, we see that $4x^2 - 21x - 18 < 0$ when

$$-\frac{3}{4} < x < 6.$$

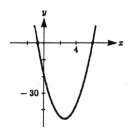

FIGURE 6.3

(b) The solution is the same as that given in part (a) except that the endpoints are included. Thus, we have $-\frac{3}{4} \le x \le 6$.

SIMILAR PROBLEMS ▬▬▬▬▬▬▬▬▬

Solve the inequalities for x.

1. $x^2 + x - 2 < 0$ **2.** $x^2 + x - 2 > 0$
3. $x^2 + 5x > 4x + 2$ **4.** $4x^2 - 21x - 18 > 0$
5 $4x^2 - 21x - 18 \geq 0$

ANSWERS

1. $-2 < x < 1$ **2.** $x < -2$ or $x > 1$
3. $x < -2$ or $x > 1$ **4.** $x < -\frac{3}{4}$ or $x > 6$
5. $x \leq -\frac{3}{4}$ or $x \geq 6$

EXAMPLE 7   **Understanding Absolute Value**

Write each of the following in a form that does not contain an absolute value sign. The symbol x represents a real number and the symbol n represents a natural number (1, 2, 3, ..., and so on).

(a) $|1.35|$ **(b)** $\left|-\frac{1}{2}\right|$ **(c)** $|x|$
(d) $|x^2|$ **(e)** $|n|$ **(f)** $|n - 2|$

SOLUTION

(a) $|1.35| = 1.35$

(b) $\left|-\frac{1}{2}\right| = \frac{1}{2}$

(c) $|x| = \begin{cases} x, & \text{if } x \geq 0 \\ -x, & \text{if } x < 0 \end{cases}$

(d) $|x^2| = x^2$ because x^2 is positive or zero for all x.

(e) $|n| = n$ because n is always positive.

(f) $|n - 2| = \begin{cases} 1 & \text{if } n = 1 \\ n - 2 & \text{if } n > 1 \end{cases}$ because $n - 2$ is nonnegative for $n > 1$

SIMILAR PROBLEMS ▬▬▬▬▬▬▬▬▬

Write each of the following in a form that does not contain an absolute value sign. The symbol x is a real number and n is a natural number.

1. $|2.86|$ **2.** $\left|-\frac{1}{4}\right|$ **3.** $|x|$
4. $|x^2 + 1|$ **5.** $|n + 1|$ **6.** $|n - 3|$

ANSWERS

1. 2.86   **2.** $\frac{1}{4}$ **3.** $\begin{cases} x, & \text{if } x \geq 0 \\ -x, & \text{if } x < 0 \end{cases}$

4. $x^2 + 1$ **5.** $|n + 1| = n + 1$ **6.** $|n - 3| = \begin{cases} 2, & \text{if } n = 1 \\ 1, & \text{if } n = 2 \\ n - 3, & \text{if } n > 2 \end{cases}$

EXAMPLE 8

Interpreting $|x - a| < b$

Fill in the blanks.

The statement $|x - 3| < 1$ is equivalent to ( ) $< x <$ ( ).

SOLUTION

As stated in the Review of Fundamentals at the beginning of this chapter

$$|x - 3| < 1 \text{ is equivalent to } 3 - 1 < x < 3 + 1$$
$$2 < x < 4.$$

SIMILAR PROBLEM

Fill in the blanks.

1. The statement $|x - 4| < 1$ is equivalent to ( ) $< x <$ ( ).

ANSWER

1. $3 < x < 5$

EXAMPLE 9

Using the Triangle Inequality

Show that if $|a - b| < s$ and $|c - b| < t$, then $|a - c| < s + t$.

SOLUTION

$$
\begin{aligned}
|a - c| &= |a - b + b - c| && \text{Add and subtract } b \\
&= |(a - b) + (b - c)| \\
&\leq |a - b| + |b - c| && \text{Triangle inequality} \\
&= |a - b| + |-(c - b)| \\
&= |a - b| + |c - b| \\
&< s + t
\end{aligned}
$$

SIMILAR PROBLEM

1. Prove that if $|a - b| < s$ and $|c - b| < t$, then $|a - c| < s + t$.

ANSWER

1. The solution is given in Example 9.

EXAMPLE 10   **Practice with ε**

Show that if $|x - 1| < \varepsilon/3$, then $|(3x + 2) - 5| < \varepsilon$.

SOLUTION

$$
\begin{aligned}
|(3x + 2) - 5| &= |3x + 2 - 5| \\
&= |3x - 3| \\
&= |3(x - 1)| \\
&= |3| \cdot |x - 1| \\
&= 3 \cdot |x - 1| \\
&< 3 \cdot \left(\frac{\varepsilon}{3}\right) \qquad \text{Use } |x - 1| < \frac{\varepsilon}{3} \\
&= \varepsilon
\end{aligned}
$$

SIMILAR PROBLEM

1. Let ε be a positive number. Show that if $|x - 1| < \varepsilon/4$, then $|(4x + 1) - 5| < \varepsilon$.

ANSWER

$$
\begin{aligned}
\textbf{1.}\quad |(4x + 1) - 5| &= |4x + 1 - 5| \\
&= |4x - 4| \\
&= |4| \, |x - 1| \\
&< 4\left(\frac{\varepsilon}{4}\right) \\
&= \varepsilon
\end{aligned}
$$

EXAMPLE 11   **More Practice with ε**

Suppose that $|x - 3| < 1$. Show that the following statements are true.

(a) $|x| > 2$ **(b)** $|x + 3| < 7$

(c) If $|x - 3| < \dfrac{\varepsilon}{7}$ for some number ε, then $|x^2 - 9| < \varepsilon$.

(d) If $|x - 3| < 6\varepsilon$ for some number ε, then $\left|\dfrac{1}{x} - \dfrac{1}{3}\right| < \varepsilon$.

SOLUTION

Recall that a statement equivalent to $|x - 3| < 1$ is

$$3 - 1 < x < 3 + 1$$
$$2 < x < 4.$$

(a) Thus, $|x| > 2$.

(b) Since

$$2 < x < 4$$
$$2 + 3 < x + 3 < 4 + 3$$
$$5 < x + 3 < 7$$

we can conclude that $|x + 3| < 7$ is true.

(c) We have

$$
\begin{aligned}
|x^2 - 9| &= |(x - 3)(x + 3)| \\
&= |x - 3||x + 3| \\
&< \left(\frac{\varepsilon}{7}\right) \cdot 7 \qquad \text{Use } |x - 3| < \frac{\varepsilon}{7} \text{ and } |x + 3| < 7 \\
&= \varepsilon.
\end{aligned}
$$

(d) We have

$$
\begin{aligned}
\left|\frac{1}{x} - \frac{1}{3}\right| &= \left|\frac{3 - x}{3x}\right| \\
&= \frac{|3 - x|}{|3x|} \\
&= \frac{|3 - x|}{|3||x|} \\
&= \frac{|3 - x|}{3|x|} \\
&< \frac{6\varepsilon}{3|x|} \qquad \text{Use } |x - 3| < 6\varepsilon \\
&= \frac{2\varepsilon}{|x|} \\
&= \frac{1}{|x|}2\varepsilon \\
&< \frac{1}{2} \cdot 2\varepsilon \qquad \text{Use } |x| > 2 \\
&= \varepsilon.
\end{aligned}
$$

SIMILAR PROBLEM

Suppose that $|x - 4| < 1$. Show that the following statements are true.

1. $|x| > 3$

2. $|x + 4| < 9$

3. If $|x - 4| < \varepsilon/9$ for a number ε, then $|x^2 - 16| < \varepsilon$

4. If $|x - 4| < 12\varepsilon$ for a number ε, then $\left|\dfrac{1}{x} - \dfrac{1}{4}\right| < \varepsilon$.

CHAPTER 6 EXERCISES

1. Are the following statements true or false?
   (a) $5 > -7$ **(b)** $-2 \geq -2$

2. Are the following statements true for all values of x?
   (a) $1/(x + 1) > 1/x$ **(b)** $-1/x \leq x$

3. Compute the following values as a function of x.
   (a) $|-0.01|$ **(b)** $|x^2 + 1|$

4. Are the following statements true or false?
   (a) $|x| = x$ **(b)** $x \leq |x|$
   (c) $-|x| \leq x$

5. Solve for x: $x^2 + x \geq 2$ **6.** Solve for x: $6x^2 - 31x + 18 \leq 0$.

7. Use the properties of absolute value to prove that $|a - b| \leq |a| + |b|$.

8. Prove that if $|x + 2| < 1$, then $-3 < x < -1$.

9. Fill in the blank with the smallest possible number.
   If $-7 < x < -2$, then $|x| < (\ \ )$.

10. Prove that if $|x + 1| < \frac{1}{3}$, then $|(3x - 9) + 12| < 1$.

11. **(a)** Prove that if $|x - 4| < 1$, then $|x + 4| < 7$.
   (b) Prove that if $|x - 2| < \varepsilon/7$, then $|x^2 - 4| < \epsilon$. Assume ε is a positive number less than 1.

12. **(a)** Prove that if $|x - 4| < 1$, then $|x| > 3$.
   (b) Prove that if $|x - 4| < 12\varepsilon$, then $\left|\dfrac{1}{x} - \dfrac{1}{4}\right| < \varepsilon$. Assume ε is a positive number less than 1.

13. Using the absolute value properties in the Chapter 6 Review of Fundamentals, prove that $|ab - cd| \leq |a|\,|b - c| + |c|\,|a - d|$.

14. **(a)** Show that if $|x - 1| < \dfrac{1}{2}$, then $|x + 4| > \dfrac{9}{4}$.
   (b) Use part (a) to show that, if ε is a positive number and $|x - 1|$ is less than both $\dfrac{1}{2}$ and $\dfrac{9\varepsilon}{4}$,
   then $\left|\dfrac{2x + 3}{x + 4} - 1\right| < \varepsilon$.

15. Solve the inequality $x^3 - x > 0$ for x. (Hint: First determine where the curve $y = x^3 - x$ intersects the x-axis.)

16. Solve: $2x^3 - 3x^2 - 2x + 3 > 0$.

17. Solve: $2x^5 + 3x^4 - 3x^3 - 2x^2 \leq 0$.

REVIEW OF FUNDAMENTALS

- A **function** is a correspondence that assigns a number, denoted $f(x)$, to every eligible number x. The formal definition of a function is: a set f of ordered pairs such that if $(x_1, y) \in f$ and $(x_2, y) \in f$, then $x_1 = x_2$.

- The **domain** of f is the set consisting of all eligible values of x. Unless otherwise specified, it consists of all numbers eligible for substitution in the formula(s) for $f(x)$.

- The **range** of f is the set of numbers resulting from the application of the function f.

- **To sketch the graph of a function** f, replace $f(x)$ by y, and sketch the graph of the resulting equation.

- **To find the domain and range from the graph.** The domain is the projection (the "shadow cast") on the x-axis by the graph, and the range is the projection on the y-axis.

EXAMPLE 1

Sketching the Graphs of Functions

Find the domains of the following functions.

(a) $f(x) = x^{20}$

(b) $g(x) = \dfrac{1}{x - 3}$

(c) $h(x) = \dfrac{1}{4x^2 - 21x - 18}$

(d) $k(x) = \sqrt{4x^2 - 21x - 18}$

(e) $p(x) = \dfrac{1}{\sqrt{4x^2 - 21x - 18}}$

SOLUTION

Since only the domains (and not the ranges) are required, and since the graphs are not easy to sketch, we shall determine the domains by describing the values of x that are **eligible** for substitution into each formula.

(a) The domain is all real numbers because all real numbers are eligible for substitution into x^{20}.

(b) Any value of x can be substituted into $1/(x - 3)$ except for the value of x that yields the meaningless expression $1/0$. This will occur when $x - 3 = 0$ or $x = 3$. Thus, the domain consists of all real numbers except 3.

(c) As in part (b), the only values of x that are not eligible for substitution into $1/(4x^2 - 21x - 18)$ are those which result in $4x^2 - 21x - 18 = 0$. This quadratic equation factors into $(4x + 3)(x - 6) = 0$, giving $4x + 3 = 0$ and $x - 6 = 0$, or $x = -3/4$ and 6. Thus, the domain consists of all real numbers except $-3/4$ and 6.

(d) The values of x that are not eligible for substitution into $\sqrt{4x^2 - 21x - 18}$ are those which cause $4x^2 - 21x - 18 < 0$, because the square root of a negative number is not real. Thus, the domain of this function is the set of all real numbers x such that $4x^2 - 21x - 18 \geq 0$. This set consists of two intervals: $(-\infty, -3/4]$ and $[6, \infty)$. (See Example 6a, page 50.)

(e) The values of x that are not eligible for substitution into $1/(\sqrt{4x^2 - 21x - 18}\,)$ are those which cause $4x^2 - 21x - 18 < 0$, because these values result in the square root of a negative number, together with those which make the denominator zero, because such values result in the meaningless $1/0$. Putting together these two conditions, we see that the domain consists of the two intervals $(-\infty, -3/4)$ and $(6, \infty)$.   ■

SIMILAR PROBLEMS

Find the domains of the following functions.

1. $q(x) = x^{18}$ **2.** $b(x) = \dfrac{1}{x - 2}$

3. $d(x) = \dfrac{1}{x^2 - x - 2}$ **4.** $g(x) = \sqrt{x^2 - x - 2}$

5. $u(x) = \dfrac{1}{\sqrt{x^2 - x - 2}}$

ANSWERS

1. All real numbers   **2.** All real numbers except 2
3. All real numbers except 2 and -1 **4.** $(-\infty, -1] \cup [2, \infty)$
5. $(-\infty, -1) \cup (2, \infty)$

EXAMPLE 2

Sketching the Graph of a Function

Sketch the graph of $f(x) = 4x - 2$ and give its domain and range.

SOLUTION

To sketch the graph of $f(x) = 4x - 2$, we sketch the graph $y = 4x - 2$. This is a straight line with intercepts $(0, -2)$ and $\left(\tfrac{1}{2}, 0\right)$. As shown in Figure 7.1, the domain of f, the projection of the graph on the x-axis, is the set of all real numbers. The range, the projection on the y-axis, consists of all real numbers.

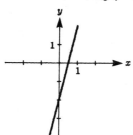

FIGURE 7.1   ■

EXAMPLE 3

Sketching the Graph of a Function

Sketch the graph of $g(x) = x$ and give its domain and range.

SOLUTION

To sketch the graph of $g(x) = x$, we sketch the graph of $y = x$. This is the straight line with slope 1 passing through the origin. As shown in Figure 7.2, the domain is all real numbers and the range is all real numbers.

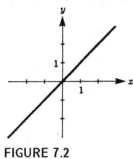

FIGURE 7.2

EXAMPLE 4

Sketching the Graph of a Function

Sketch the graph of $h(x) = 3$ and give its domain and range.

SOLUTION

To sketch the graph of $h(x) = 3$, we sketch the graph of $y = 3$. This is the horizontal line through the coordinate 3 on the y-axis. (See Figure 7.3.) Again, the domain is the set of all real numbers. The range, the projection of the graph onto the y-axis, consists simply of the number 3.

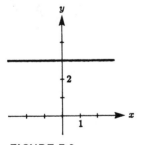

FIGURE 7.3

EXAMPLE 5

Sketching the Graph of a Function

Sketch the graph of $k(x) = c$ and give its domain and range.

SOLUTION

To sketch the graph of $k(x) = c$, we sketch the graph of $y = c$. This is the horizontal line through c on the y-axis. The range is c and the domain is all real numbers.

EXAMPLE 6

Sketching the Graph of a Function

Sketch the graph of $p(x) = 4 - x^2$ and give its domain and range.

SOLUTION

Replacing $p(x)$ by y in the equation $p(x) = 4 - x^2$, we obtain $y = 4 - x^2$.
(See Figure 7.4.)

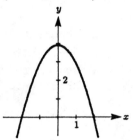

FIGURE 7.4

The graph of this function is a parabola. The domain, the projection on the x-axis, consists of all real numbers. The range, the projection on the y-axis, consists of $(-\infty, 4]$.

EXAMPLE 7

Sketching the Graph of a Function

Sketch the graph of $g(x) = 4 - x^2$, $1 \le x < 3$ and give its domain and range.

SOLUTION

The domain of $g(x)$ is specified as $[1, 3)$. To sketch the graph, first ignore this restriction on the domain and graph $y = 4 - x^2$ as in Example 6. (See Figure 7.5.) Now indicate the portion of the graph that has $[1, 3)$ as its domain. This will be the graph of g. The range of g is the projection on the y-axis, which is $(-5, 3]$.

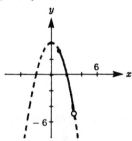

FIGURE 7.5

SIMILAR PROBLEMS ▮▮▮▮▮▮▮▮▮▮▮

Sketch the graphs of the following functions, and give their domains and ranges.

1. $f(x) = 2x - 3$ **2.** $g(x) = 2x$
3. $h(x) = 5$ **4.** $p(x) = 9 - x^2$
5. $q(x) = 9 - x^2, 1 \le x < 3$

ANSWERS

1. Domain: $(-\infty, \infty)$ **2.** Domain: $(-\infty, \infty)$
    Range: $(-\infty, \infty)$ Range: $(-\infty, \infty)$

3. Domain: $(-\infty, \infty)$ **4.** Domain: $(-\infty, \infty)$
    Range: $\{5\}$ Range: $(-\infty, 9]$

5. Domain: $[1, 3)$
    Range: $(0, 8]$

EXAMPLE 8 ▮▮▮▮

Evaluating a Function

For $g(x) = x^3 - 7x^2 + x + 1$, find $g(x + h)$.

SOLUTION

To find $g(x + h)$, we must substitute $x + h$ for x in the formula for $g(x)$.

$$g(x) = x^3 - 7x^2 + x + 1$$
$$g(x + h) = (x + h)^3 - 7(x + h)^2 + (x + h) + 1$$
$$= x^3 + 3x^2h + 3xh^2 + h^3 - 7(x^2 + 2xh + h^2) + (x + h) + 1$$
$$= x^3 + 3x^2h + 3xh^2 + h^3 - 7x^2 - 14xh - 7h^2 + x + h + 1$$

▮▮▮▮

EXAMPLE 9 ▮▮▮▮

Evaluating a Function

For $k(x) = 3$, find $k(7)$ and $k(x + h)$.

SOLUTION

The function $k(x)$ is a constant function because its value is always 3, thus
$k(7) = 3$ and $k(x + h) = 3$. ▮▮▮▮

EXAMPLE 10   **Forming a Difference Quotient**

For $f(x) = \dfrac{1}{x}$, find

$$\frac{f(x+h) - f(x)}{h}.$$

SOLUTION

For $f(x) = 1/x$, we must have $f(x+h) = 1/(x+h)$.

$$
\begin{aligned}
\frac{f(x+h) - f(x)}{h} &= \frac{\dfrac{1}{x+h} - \dfrac{1}{x}}{h} \\
&= \frac{\dfrac{x - (x+h)}{(x+h)x}}{h} \\
&= \frac{\dfrac{-h}{(x+h)x}}{h} \\
&= \frac{-h}{(x+h)xh} \\
&= \frac{-1}{(x+h)x}
\end{aligned}
$$

SIMILAR PROBLEMS

1. For $g(x) = 2x^3 + 3x^2 - x + 4$, find $g(x+h)$.
2. For $k(x) = 5$, find $k(2)$ and $k(x+h)$.
3. For $f(x) = \dfrac{2}{x}$, find $\dfrac{f(x+h) - f(x)}{h}$.

ANSWERS

1. $2x^3 + 6x^2h + 6xh^2 + 2h^3 + 3x^2 + 6xh + 3h^2 - x - h + 4$
2. $k(2) = 5, \quad k(x+h) = 5$
3. $\dfrac{-2}{x(x+h)}$

EXAMPLE 11   **Interpreting $f(x)$ Graphically**

Sketch the graph of $f(x) = x^2/4$. Let h be a positive number, and let x be in the domain of f. Indicate the following on the graph of f.

(a) $f\left(\dfrac{3}{2}\right)$ **(b)** $f\left(\dfrac{3}{2} + h\right)$

SOLUTION

To sketch the graph of $f(x) = x^2/4$, we sketch $y = x^2/4$. This is a parabola whose only intercept is $(0, 0)$. (See Figure 7.6.)

(a) To locate $f\left(\frac{3}{2}\right)$ on the graph, we first locate the x-coordinate $\frac{3}{2}$ on the x-axis. Then $f\left(\frac{3}{2}\right) = \frac{9}{16}$ is the y-coordinate. The quantity $f\left(\frac{3}{2}\right)$ can also be seen as the vertical distance from the x-coordinate $\frac{3}{2}$ to the curve since the curve is above the x-axis.

(b) Since h is a positive number, we locate $\frac{3}{2} + h$ to the right of $\frac{3}{2}$. Then $f\left(\frac{3}{2} + h\right)$ is the distance from the x-axis to the curve.

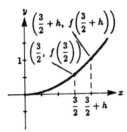

FIGURE 7.6

SIMILAR PROBLEMS

Sketch the graph of $f(x) = x^2/3$. Let h be a positive number, and let x be in the domain of f. Indicate the following on the graph of f.

1. $f(2)$ **2.** $f(2 + h)$

ANSWERS

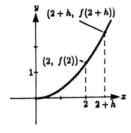

EXAMPLE 12

Sketching the Graph of a Function

Sketch the graph of the function.

$$f(x) = \begin{cases} 1, & \text{if } x \leq 0 \\ -1, & \text{if } x > 0 \end{cases}$$

SOLUTION

We first lightly sketch $y = 1$ and $y = -1$. Next, we fill in the portion of $y = 1$ corresponding to $x \leq 0$. Finally, we fill in the portion of $y = -1$ corresponding to $x > 0$. The result is Figure 7.7.

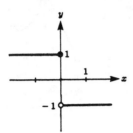

FIGURE 7.7

EXAMPLE 13

Sketching the Graph of a Function

Sketch the graph of the function.

$$g(x) = \begin{cases} 2x, & \text{if } x < -1 \\ 2x^2 + x - 3, & \text{if } -1 \leq x < 2 \\ -x + 3, & \text{if } x \geq 2 \end{cases}$$

SOLUTION

To sketch the graph of g, we first lightly sketch the three formulas: $y = 2x$ (straight line through the origin with slope 2), $y = 2x^2 + x - 3$ (parabola with x-intercepts at $(-3/2, 0)$ and $(1, 0)$), and $y = -x + 3$ (straight line with intercepts $(0, 3)$ and $(3, 0)$). Next, we fill in the portions of these three graphs that are indicated by limitations on x. The graph of $y = -x + 3$ is filled in for $x \geq 2$. Substitute $x = 2$ into the formula $y = -x + 3$. This portion of the graph ends at the point $(2, 1)$. The endpoints of the $y = 2x^2 + x - 3$ portion are $(2, 7)$ and $(-1, -2)$. We obtain this by substituting $x = 2$ and $x = -1$ into the formula $y = 2x^2 + x - 3$. The point $(-1, -2)$ is also the endpoint of the $y = 2x$ portion, so that the curve is continuous at that point, as shown in Figure 7.8.

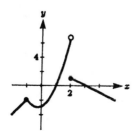

FIGURE 7.8

SIMILAR PROBLEMS ▰▰▰▰▰▰▰▰▰

Sketch the graphs of the functions.

1. $f(x) = \begin{cases} 2, & \text{if } x \leq 0 \\ -1, & \text{if } x > 0 \end{cases}$

2. $f(x) = \begin{cases} 2x + 4, & \text{if } x \leq -1 \\ x^2 - x, & \text{if } -1 < x \leq 1 \\ 2 - x, & \text{if } x > 1 \end{cases}$

ANSWERS

1.

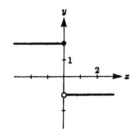

2.

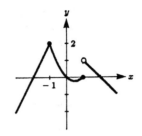

EXAMPLE 14
▰▰▰▰▰▰

Sketching the Graph of a Function

Sketch the graph of $f(x) = \sqrt{4 - x^2}$.

SOLUTION

Replacing $f(x)$ by y, we sketch the graph of $y = \sqrt{4 - x^2}$.

$$y = \sqrt{4 - x^2}$$
$$y^2 = 4 - x^2$$
$$x^2 + y^2 = 4$$
$$x^2 + y^2 = 2^2$$

This is the equation of a circle with radius 2 and center at the origin. This circle is sketched in Figure 7.9. From the formula, we can see that the function has only positive values because the square root denotes a positive number. Thus, the graph of the function consists only of the upper half of the circle as shown.

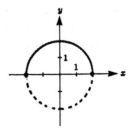

FIGURE 7.9 ▰▰▰▰

SIMILAR PROBLEM ▬▬▬▬▬▬▬

1. Sketch the graph of $f(x) = \sqrt{9 - x^2}$.

ANSWER

1.

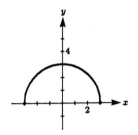

EXAMPLE 15

Writing a Formula for a Function

Obtain the formula for the function whose graph is the straight line passing through $(1, -3)$ and $(-2, -4)$.

SOLUTION

Use the point-slope formula $\left(\text{with } \dfrac{y_2 - y_1}{x_2 - x_1} \text{ as slope}\right)$

$$y - y_1 = \frac{y_2 - y_1}{x_2 - x_1}(x - x_1)$$

to write the equation of the straight line passing through $(1, -3)$ and $(-2, -4)$.

$$y - (-3) = \frac{(-4) - (-3)}{(-2) - 1}(x - 1)$$

$$y + 3 = \frac{-1}{-3}(x - 1)$$

$$y + 3 = \frac{x}{3} - \frac{1}{3}$$

$$y = \frac{x}{3} - \frac{1}{3} - 3$$

$$y = \frac{x}{3} - \frac{10}{3}$$

$$f(x) = \frac{x}{3} - \frac{10}{3}$$

▬▬▬

SIMILAR PROBLEM ▬▬▬▬▬▬▬

1. Let f be the function whose graph is the straight line passing through $(1, 3)$ and $(-6, -5)$. Find a formula for $f(x)$.

ANSWER

1. $f(x) = \dfrac{8}{7}x + \dfrac{13}{7}$

CHAPTER 7 EXERCISES

In Exercises 1–3, sketch the graph of the function and give its domain and range.

1. $f(x) = 2 - x$

2. $g(x) = x^2 - 4$

3. $k(x) = x^2 - 1, \quad -2 < x \le -1, \text{ and } \frac{1}{2} < x < 2$

4. Sketch the graph of the function.
$$f(x) = \begin{cases} -1, & \text{for } x \le 3 \\ -4, & \text{for } x > 3 \end{cases}$$

5. For $f(x) = 1/x^2$, find $\dfrac{f(x+h) - f(x)}{h}$.

6. For $k(x) = -2$, find $k(x+h)$.

7. For $f(x) = \dfrac{x^3 - 2x}{x+1}$, find $f(x+h)$.

In Exercises 8–11, find the domain of the function.

8. $f(x) = \dfrac{11}{x+5}$

9. $f(x) = \dfrac{1}{x^2 + 1}$

10. $f(x) = \dfrac{x}{x^2 + x - 2}$

11. $f(x) = \dfrac{x}{\sqrt{2 - x - x^2}}$

12. Find $f(x)$ for the function f whose graph is the straight line with slope -2 passing through the point $(1, -5)$.

13. Find the function whose graph is the straight line that passes through the point $(-1, -2)$ and intersects the x-axis at 3.

14. Sketch the graph of the function $f(x) = \dfrac{1}{2}\sqrt{36 - 9x^2}$.

15. Find the maximum value of the function $f(x) = 4 - x^2$.

16. Find the minimum value of the function $f(x) = 3x - x^2$ on the closed interval $[-1, 1]$.

17. Does the function $f(x) = x + 1$ attain a maximum value on the open interval $(1, 3)$?

18. Sketch the graph of the function f.
$$f(x) = \begin{cases} -2x, & x \le -2 \\ x^2 - x - 2, & -2 < x < 2 \\ -\frac{1}{2}x, & x \ge 2 \end{cases}$$

19. Sketch the graph of the function f.
$$f(x) = \begin{cases} x^2 + 3x + 2, & \text{for } x \le 0 \\ x^2 - 3x + 2, & \text{for } x \ge 0 \end{cases}$$

20. Sketch the graph of $f(x) = x^2 - x$ and $g(x) = 2 - 2x$. Find the values of x for which $g(x) \ge f(x)$.

21. Show that the equations $f(x+t) = f(x) + f(t)$ and $g(x+h) = g(x) + h$ are not valid for all functions f and g. Produce examples of f and g for which the equations are valid.

8

Functions II: Combinations of Functions, Difficult Graphs

REVIEW OF FUNDAMENTALS

- The **sum, product, and quotient of functions** f and g are defined by the following statements.

  (a) $(f + g)(x) = f(x) + g(x)$

  (b) $(fg)(x) = f(x) \cdot g(x)$

  (c) $\left(\dfrac{f}{g}\right)(x) = \dfrac{f(x)}{g(x)}, \quad g(x) \neq 0$

- The **composition of functions** f and g is defined by $(f \circ g)(x) = f(g(x))$. Note that the domain of $f \circ g$ is $\{x \ : \ x \in \text{domain of } g \text{ and } g(x) \in \text{domain of } f\}$.

- The **inverse of a function** f, denoted by f^{-1} has the following properties.

  (a) The inverse of f *exists* if no horizontal line intersects the graph of f in two or more places. Functions of this kind are called one-to-one. Formally, a function is one-to-one if $f(x_1) = f(x_2)$ implies $x_1 = x_2$.

  (b) The inverse of f is *expressed* in terms of x and y by solving the original equation $y = f(x)$ for x in terms of y, and then interchanging the symbols x and y.

  (c) The inverse of f *has as its domain* the range of f, and *has as its range* the domain of f.

  (d) The inverse of f *is graphed* by reflecting the graph of the original function about the line $y = x$.

EXAMPLE 1

Sums, Products, and Quotients of Functions

Let $f(x) = x^2 + 3$, $g(x) = 1/x$, and $k(x) = 2$.

Find **(a)** $(f + g)(x)$, **(b)** $(fg)(x)$, and **(c)** $(k/f)(x)$.

SOLUTION

(a) $(f + g)(x) = f(x) + g(x) = (x^2 + 3) + \dfrac{1}{x} = x^2 + \dfrac{1}{x} + 3$

(b) $(fg)(x) = [f(x)][g(x)] = [x^2 + 3]\left[\dfrac{1}{x}\right] = x + \dfrac{3}{x}$

(c) $\left(\dfrac{k}{f}\right)(x) = \dfrac{k(x)}{f(x)} = \dfrac{2}{x^2 + 3}$

EXAMPLE 2   **Sums of Functions**

Let $f(x) = x^2 + 3$ and $k(x) = 2$. Express the function $q(x) = x^2 + 5$ as the sum of the two given functions.

SOLUTION

$$q(x) = x^2 + 5 = (x^2 + 3) + 2 = f(x) + k(x)$$

Thus, $q = f + k$.

EXAMPLE 3   **Graphing a Sum of Functions**

Sketch the graph of the function $p(x) = (1/x) + 2$ by first sketching the graph of $g(x) = 1/x$. Observe *the effect of adding a constant function.*

SOLUTION

First we graph $g(x) = 1/x$, as shown in Figure 8.1(a). For each ordinate x, the point on the graph of $p(x) = (1/x) + 2$ is 2 units higher than the point on $g(x) = 1/x$. Thus, to obtain the graph of

$$p(x) = \frac{1}{x} + 2$$

we raise the entire graph of $g(x) = 1/x$ by 2 units, as shown in Figure 8.1(b). The effect of adding a constant function, therefore, is to raise (or lower) the original graph by the constant amount.

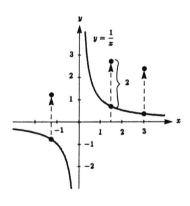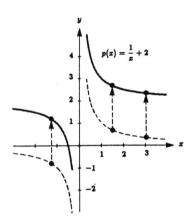

FIGURE 8.1

SIMILAR PROBLEMS

Let $f(x) = x^2 - 1$, $g(x) = 1/x$, and $k(x) = 3$.

1. Find $(f + g)(x)$, $(fg)(x)$, and $(k/f)(x)$.

2. Express the function $q(x) = x^2 + 2$ as the sum of two of the given functions.

3. Sketch the graph of the function $p(x) = (1/x) + 1$.

ANSWERS

1. $x^2 + \dfrac{1}{x} - 1$, $\quad x - \dfrac{1}{x}$, $\quad \dfrac{3}{x^2 - 1}$

   2. $q = f + k$

3.

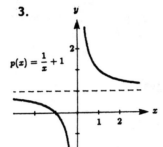

$p(x) = \dfrac{1}{x} + 1$

EXAMPLE 4

Composition of Functions

For $f(x) = x^2 + 3$ and $g(x) = 1/x$, find the following.

(a) $(f \circ g)(x)$ **(b)** $(g \circ f)(x)$

SOLUTION

(a) $(f \circ g)(x) = f(g(x)) = f\left(\dfrac{1}{x}\right) = \left(\dfrac{1}{x}\right)^2 + 3 = \dfrac{1}{x^2} + 3$

(b) $(g \circ f)(x) = g(f(x)) = g(x^2 + 3) = \dfrac{1}{x^2 + 3}$

EXAMPLE 5

Composition of Functions

Express $h(x) = |2x - 4|$ as a composition of simpler functions.

SOLUTION

We can express the function $h(x) = |2x-4|$ as the absolute value function applied to the function $2x-4$. Thus, if we let $a(x) = |x|$ and $p(x) = 2x-4$, then $h = a \circ p$.

CHECK $(a \circ p)(x) = a(p(x)) = a(2x - 4) = |2x - 4| = h(x)$

EXAMPLE 6   **Composition of Functions**

Express the following functions as compositions of simpler functions.

(a) $k(x) = \sqrt{3-x}$

(b) $b(x) = \sqrt{x^2 - 2x}$

SOLUTION

(a) We can express the function

$$k(x) = \sqrt{3-x}$$

as the square root function applied to the function $3 - x$. Thus, if we let $s(x) = \sqrt{x}$ and $q(x) = 3 - x$, then $k = s \circ q$.

CHECK

$$(s \circ q)(x) = s(q(x))$$
$$= s(3 - x)$$
$$= \sqrt{3 - x}$$
$$= k(x)$$

(b) Similarly, the function

$$b(x) = \sqrt{x^2 - 2x}$$

can be expressed as $b = s \circ r$, where $r(x) = x^2 - 2x$.

SIMILAR PROBLEMS

1. For $f(x) = x^2 - 1$ and $g(x) = 1/x$, find $(f \circ g)(x)$ and $(g \circ f)(x)$.

2. Express $h(x) = |3x + 1|$ as a composition of simpler functions.

3. Express $k(x) = \sqrt{2 + x}$ and $b(x) = \sqrt{x^2 + 3}$ as compositions of simpler functions.

ANSWERS

1. $\dfrac{1}{x^2} - 1$ and $\dfrac{1}{x^2 - 1}$

2. $h = f \circ g$ where $f(x) = |x|$ and $g(x) = 3x + 1$

3. $k = s \circ t$ and $b = s \circ q$ where $s(x) = \sqrt{x}$, $t(x) = 2 + x$, and $q(x) = x^2 + 3$

EXAMPLE 7

Absolute Value Function and Graphing

Graph the following functions.

(a) $f(x) = |x|$ **(b)** $f(x) = |2x - 4|$

SOLUTION

Both functions to be graphed are of the form $|p(x)|$, which is the composition of the absolute value function and the function $p(x)$. We use this composite structure to graph $|p(x)|$ by first graphing the inner function $p(x)$, and then reflecting about the x-axis the parts that are below the x-axis.

(a) To graph the function $f(x) = |x|$, first sketch $y = x$, then reflect the part below the x-axis, as shown in Figure 8.2.

(b) To graph the function $g(x) = |2x - 4|$, first sketch $y = 2x - 4$, then, reflect the part below the x-axis as shown in Figure 8.3.

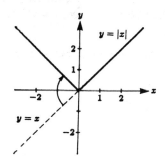

FIGURE 8.2   FIGURE 8.3

SIMILAR PROBLEM

Sketch the graph of $g(x) = |2x + 1|$.

ANSWER

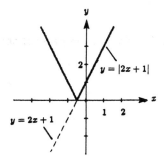

EXAMPLE 8   **Inverse Functions**

Let $f(x) = 2x - 2$.

(a) Sketch the graph of f.

(b) Determine whether f has an inverse function.

(c) Sketch the graph of f^{-1}.

(d) Give a formula for $f^{-1}(x)$.

SOLUTION

(a) The graph of $f(x) = 2x - 2$ is shown in Figure 8.4. Following the procedure in Chapter 1, we obtain the intercepts $(0, -2)$ and $(1, 0)$.

(b) The function f *does* have an inverse because no horizontal line intersects its graph more than once.

(c) The graph of the inverse is obtained by reflecting the original graph about the line $y = x$ (as if the line $y = x$ were a mirror). In this example, a straight line is obtained.

(d) To obtain the formula for the inverse, *first* solve the original equation for x.

$$y = 2x - 2$$
$$2 + y = 2x$$
$$\frac{2+y}{2} = x$$
$$\frac{y}{2} + 1 = x$$
$$x = \frac{y}{2} + 1$$

Then interchanging x and y, we obtain

$$y = \frac{x}{2} + 1.$$

Thus, the inverse is

$$f^{-1}(x) = \frac{x}{2} + 1.$$

Observe that the graph of f^{-1} obtained in part (c) agrees with this formula.

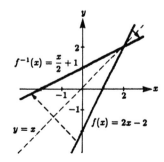

FIGURE 8.4

EXAMPLE 9   **Inverse Functions**

Let $q(x) = 4 - x^2$, $1 \le x < 3$.

(a) Sketch the graph of q.

(b) Determine whether q has an inverse function.

(c) Sketch the graph of q^{-1}.

(d) Give a formula for $q^{-1}(x)$.

(e) Find the domain and range of q^{-1}.

SOLUTION

(a) The graph of $q(x) = 4 - x^2$, $1 \le x < 3$ is obtained by first sketching $y = 4 - x^2$, as shown in Figure 8.5, and then restricting the domain to $[1, 3)$.

(b) The function q *does* have an inverse because no horizontal line intersects its graph more than once. Note that this is not true for the function $g(x) = 4 - x^2$ without the restriction on x. Many horizontal lines intersect its graph twice.

(c) The graph of the inverse is obtained by reflecting the original graph about the line $y = x$, as shown in Figure 8.5.

(d) To obtain the formula for the inverse, *first* solve the original equation for x.

$$y = 4 - x^2, \qquad 1 \le x < 3 \quad \text{and} \quad -5 < y \le 3$$
$$x^2 + y = 4$$
$$x^2 = 4 - y$$
$$x = \sqrt{4 - y}$$

Note that since x must be positive, we have $x = \sqrt{4 - y}$, not $x = -\sqrt{4 - y}$ or $x = \pm\sqrt{4 - y}$. Then interchanging x and y, we obtain

$$y = \sqrt{4 - x}, \qquad -5 < x \le 3.$$

Thus, the inverse is

$$q^{-1}(x) = \sqrt{4 - x}, \qquad -5 < x \le 3.$$

(e) The domain of q^{-1} is the range of q, which is $(-5, 3]$. The range of q^{-1} is the domain of q, which is $[1, 3)$.

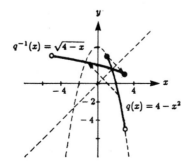

FIGURE 8.5

SIMILAR PROBLEMS

1. Sketch the graph of $f(x) = \frac{1}{2}x - 1$, check that f has an inverse, graph $f^{-1}(x)$, and give a formula for $f^{-1}(x)$.

2. Sketch the graph of $q(x) = 9 - x^2, 1 \leq x < 4$, check that q has an inverse, graph $q^{-1}(x)$, and give a formula for $q^{-1}(x)$. Find the domain and range of q^{-1}.

ANSWERS

1. $f^{-1}(x) = 2x + 2$

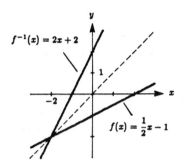

2. $q^{-1}(x) = \sqrt{9-x}$

    Domain: $(-7, \ 8]$
    Range: $[1, \ 4)$

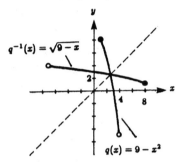

EXAMPLE 10

Graphing a Function Using an Inverse

Sketch the graph of $f(x) = \sqrt{x}$.

SOLUTION

We begin by determining that $f(x) = \sqrt{x}$ is the inverse function of $p(x) = x^2$, $0 \leq x$. We know this because, if $y = \sqrt{x}$, then $x = y^2$. First sketch $y = x^2$, then reflect the graph about the line $y = x$. Since the domain and range of $\sqrt{x}$ are $[0, \ \infty)$, we reflect only the part of $y = x^2$ lying in the first quadrant, as shown in Figure 8.6.

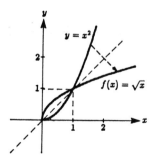

FIGURE 8.6

EXAMPLE 11　　**Graphing a Function Using an Inverse**

Sketch the graph of $g(x) = x^{1/3}$.

SOLUTION

The function $g(x) = x^{1/3}$ is the inverse of the function $q(x) = x^3$ (because, if $y = x^{1/3}$, then $x = y^3$). So we first sketch $y = x^3$, and then reflect its graph about the line $y = x$, as shown in Figure 8.7.

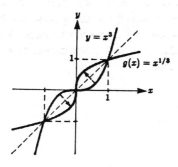

FIGURE 8.7

SIMILAR PROBLEMS

1. Sketch the graph of $f(x) = \sqrt[4]{x}$.

2. Sketch the graph of $g(x) = x^{1/5}$.

ANSWERS

1.
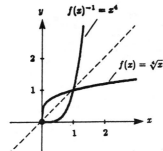

2.
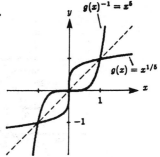

EXAMPLE 12   **Working with ϵ and δ**

Graph the function $f(x) = 2x$. Find a number δ that satisfies the following condition.

$$\text{Whenever } |x - 1| < \delta, \ |f(x) - 2| < 1.$$

SOLUTION

The graph of $y = 2x$ (a straight line with slope 2 and y-intercept 0) is shown in Figure 8.8. In Chapter 6, we saw that we must find a number δ that satisfies the statement "whenever x is in $(1 - \delta, \ 1 + \delta)$, $f(x)$ is in $(2 - 1, \ 2 + 1)$." This latter open interval, $(1, 3)$, is sketched on the y-axis, as shown in Figure 8.9. We must find an open interval on the x-axis with center 1 and radius δ that is small enough so that every value of x in it corresponds to a value of $f(x)$ in the open interval $(1, 3)$ on the y-axis.

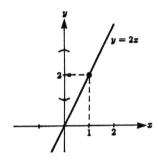

FIGURE 8.8

By tracing from $(1, 3)$ on the y-axis back to the x-axis, we can see that every value of x in the open interval $\left(1 - \frac{1}{2}, \ 1 + \frac{1}{2}\right)$ corresponds to a value of $f(x)$ in $(1, 3)$. Thus, $\frac{1}{2}$ is an acceptable value for δ. (Any positive value smaller than $\frac{1}{2}$ also would be acceptable.)

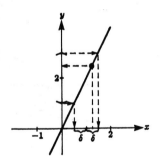

FIGURE 8.9

EXAMPLE 13

Working with ϵ and δ

Using the graph of $g(x) = \sqrt{x}$, find any number δ that satisfies the following condition.

$$\text{Whenever } |x - 4| < \delta, \quad |g(x) - 2| < 1.$$

SOLUTION

The graph of $y = \sqrt{x}$ (obtained by graphing $y = x^2$ and reflecting) is shown in Figure 8.10(a). As in Example 12, we must find an open interval on the x-axis with center 4 and radius δ that is small enough so that every value of x in this interval corresponds to a value of $g(x)$ in the interval $(1, 3)$ on the y-axis.

By tracing back from $(1, 3)$ on the y-axis, we can see that every value of x in the interval $(1, 7)$ corresponds to a value of $f(x)$ in $(1, 3)$. [See Figure 8.10(b).] Notice that the interval $(1, 7)$ is chosen with a center at 4 so as to be of the form $(4 - \delta, 4 + \delta)$. Therefore, $\delta = 3$ is an acceptable value.

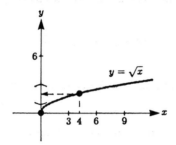 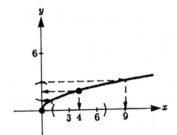

FIGURE 8.10

SIMILAR PROBLEMS

1. Graph the function $f(x) = 3x$. Find any number δ that satisfies the statement: "Whenever $|x - 1| < \delta$, $|f(x) - 3| < 1$."

2. Using the graph of $g(x) = \sqrt{x}$, find any number δ that satisfies the condition: "Whenever $|x - 9| < \delta$, $|g(x) - 3| < 1$."

ANSWERS

1. $\dfrac{1}{3}$ or smaller positive number

2. 5 or smaller positive number

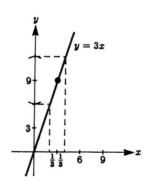 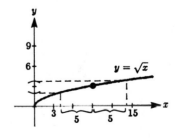

CHAPTER 8 EXERCISES

1. Sketch the graph of the function $f(x) = |-2x - 1|$.

2. For $f(x) = \sqrt{x^2 + 1}$ and $g(x) = 1/x$, find the following.
   (a) $(f + g)(x)$ (b) $(fg)(x)$
   (c) $(f \circ g)(x)$ (d) $(g \circ f)(x)$

3. Express each of the following functions as the composition of two simpler functions.
   (a) $f(x) = |x^2 - 1|$ (b) $g(x) = \sqrt{x^3 - 1}$
   (c) $k(x) = (4x - 5)^{3/4}$

4. Express the following function as a composition of simpler functions.

$$f(x) = \frac{5}{\sqrt{2x + 1}} - 3(2x + 1)^2$$

5. Graph $f(x) = |x - 1|$. Use this graph and the sum-of-functions concept to graph $k(x) = |x - 1| + 1$.

6. Graph the following functions.
   (a) $f(x) = \sqrt{-2x - 1}$ (b) $f(x) = \sqrt{9 - x^2}$
   (c) $f(x) = \sqrt{2 - x^2}$

7. For $f(x) = 2 - 3x$, graph $f^{-1}(x)$ using the graph of f, and obtain a formula for $f^{-1}(x)$.

8. Sketch the inverse of the function $f(x) = x^2 + x - 2$, $x \geq 0$.

9. Sketch the graphs of the following functions by using the inverse function concept.
   (a) $f(x) = x^{1/16}$ (b) $g(x) = x^{1/7}$

10. Are $f(x)^{-1}$ and $f^{-1}(x)$ equal for all functions and all values of x? If not, give an example of a function for which the two are not equal.

9

Simultaneous Equations

REVIEW OF FUNDAMENTALS

■ **Methods for Solving Simultaneous Equations**

1. **Substitution**

   (a) From one of the equations, solve for one unknown in terms of the other(s).

   (b) Substitute into the other equation(s).

2. **Linear equation method**

3. A combination of methods (1) and (2).

■ **Intersecting Curves**

To find the coordinates of the points at which two curves intersect, solve the equations of the two curves simultaneously.

EXAMPLE 1

Solving Simultaneous Equations

Solve the following simultaneous linear equations for x and y.

$$3x - 2y = 7, \qquad 4x + 5y = -6$$

SOLUTION

Since both equations are linear, it is usually easiest to eliminate one of the unknowns by the following method.

Eliminate one unknown. In this example, we eliminate y as follows.

$3x - 2y = 7$	Equation 1
$4x + 5y = -6$	Equation 2
$15x - 10y = 35$	Multiply Equation 1 by 5
$8x + 10y = -12$	Multiply Equation 2 by 2
$23x = 23$	Add
$x = 1$	Solve for x

Substitute in one of the original equations.

$3x - 2y = 7$	Substitute $x = 1$
$3(1) - 2y = 7$	
$-2y = 7 - 3$	
$-2y = 4$	
$y = -2$	

Thus, the solution is $x = 1$ and $y = -2$.

SIMILAR PROBLEM ▬▬▬▬▬▬▬▬▬▬▬▬▬▬▬

1. Solve the equations simultaneously: $2x + 3y = 1$, $5x - 4y = -9$

ANSWER

1. $x = -1$ and $y = 1$.

EXAMPLE 2 ▬▬▬▬▬▬   **Solving Simultaneous Equations**

Solve for x and y in the following pair of equations.

$$y - x + 1 = 0 \qquad \text{Equation 1}$$
$$2y = 3x - 3x^2 + y^2 \qquad \text{Equation 2}$$

SOLUTION

Since one equation contains squares, the substitution method must be used. From the most convenient equation, solve for one unknown in terms of the other.

$$y - x + 1 = 0 \qquad \text{Equation 1}$$
$$y = x - 1$$

Substitute for y in the other equation and solve for x.

$$2y = 3x - 3x^2 + y^2$$
$$2(x - 1) = 3x - 3x^2 + (x - 1)^2$$
$$2x - 2 = 3x - 3x^2 + (x^2 - 2x + 1)$$
$$2x - 2 = 3x - 3x^2 + x^2 - 2x + 1$$
$$2x - 2 = 1 + x - 2x^2$$
$$2x^2 + x - 3 = 0$$
$$(2x + 3)(x - 1) = 0$$
$$x = -\frac{3}{2}, \ 1$$

Substitute *each* solution into the simpler of the two given equations: Equation 1.

When $x = -\dfrac{3}{2}$, $y = x - 1 = -\dfrac{3}{2} - 1 = -\dfrac{5}{2}$.

When $x = 1$, $y = x - 1 = 1 - 1 = 0$.

Thus, one solution is $x = -\frac{3}{2}$ and $y = -\frac{5}{2}$ and the other solution is $x = 1$ and $y = 0$. ▬▬▬▬▬

SIMILAR PROBLEM ▰▰▰▰▰▰▰▰▰▰▰▰▰

1. Solve for x and y: $y + x - 3 = 0$, $2y = 6x + 2x^2 - y^2$

ANSWER

1. $x = 1$ and $y = 2$, $x = -15$ and $y = 18$

EXAMPLE 3
▰▰▰▰▰▰▰

Solving Simultaneous Equations

Solve for A, B, C, and D.

$$A + C = 0 \qquad \text{Equation 1}$$
$$B - D = 0 \qquad \text{Equation 2}$$
$$4A + 3D + 2C = 7 \qquad \text{Equation 3}$$
$$A + 4B + 6C = -6 \qquad \text{Equation 4}$$

SOLUTION

The objective is to obtain (if possible) a pair of equations containing a total of only two unknowns. Note that eliminating both A and D from Equation 3 and Equation 4 in the present example would have the desired effect. The result would be two equations in B and C only. To do this, use the simplest equations to obtain one unknown in terms of the other. From the first two equations, we conclude that $A = -C$ and $D = B$. Substituting these values into Equation 3 produces the following.

$$4A + 3D + 2C = 7 \qquad \text{Equation 3}$$
$$4(-C) + 3B + 2C = 7 \qquad \text{Substitute } A = -C \text{ and } D = B$$
$$-4C + 3B + 2C = 7$$
$$3B - 2C = 7 \qquad \text{Equation 5}$$

Substituting $A = -C$ into Equation 4 produces the following.

$$A + 4B + 6C = -6 \qquad \text{Equation 4}$$
$$(-C) + 4B + 6C = -6 \qquad \text{Substitute } A = -C$$
$$4B + 5C = -6 \qquad \text{Equation 6}$$

Now we solve Equations 5 and 6 as follows.

$$3B - 2C = 7 \qquad \text{Equation 5}$$
$$4B + 5C = -6 \qquad \text{Equation 6}$$
$$15B - 10C = 35 \qquad \text{Multiply Equation 5 by 5}$$
$$8B + 10C = -12 \qquad \text{Multiply Equation 6 by 2}$$
$$23B = 23$$
$$B = 1$$

Since $B = 1$, we can conclude from Equation 5 that $C = -2$. Also, $A = -C = -(-2) = 2$ and $D = B = 1$. Thus, the solution is $A = 2$, $B = 1$, $C = -2$, and $D = 1$.   ■

SIMILAR PROBLEM

1. Solve the simultaneous equation for P, Q, S, and T.

$$P + Q = 0, \quad S + T = 1$$
$$3P + 2T + 5Q = -10$$
$$P + S + 2Q = 2$$

ANSWER

1. $P = 2$, $Q = -2$, $S = 4$, $T = -3$

EXAMPLE 4

Finding the Points of Intersection of Graphs

(a) Find the points at which the straight lines $3x - 2y = 7$ and $4x + 5y = -6$ intersect.

(b) Find the x-coordinates of the points at which the curves $y - x + 1 = 0$ and $2y = 3x - 3x^2 + y^2$ intersect.

SOLUTION

(a) If two lines intersect at a point (x, y) then (x, y) lies on both lines. Thus, (x, y) satisfies both equations, and so we must solve the two equations simultaneously. To solve $3x - 2y = 7$ and $4x + 5y = -6$ simultaneously, see Example 1. We obtain $x = 1$ and $y = -2$. Thus, the point of intersection is $(1, -2)$.

(b) To find the points at which the curves $y - x + 1 = 0$ and $2y = 3x - 3x^2 + y^2$ intersect, we must solve these two equations simultaneously. (See Example 2.) The x-coordinates of the points found in Example 2 are 1 and $-\frac{3}{2}$.   ■

SIMILAR PROBLEM

1. Find the point at which the straight lines $2x + 3y = 1$ and $5x - 4y = -9$ intersect.

2. Find the x-coordinates of the points at which the straight line $y + x - 3 = 0$ intersects the curve $2y = 6x + 2x^2 - y^2$.

ANSWER

1. $x = -1$, $y = 1$

2. 1 and -15

CHAPTER 9 EXERCISES

1. Solve the equations simultaneously: $2x + 8 = 5y$, $3y - 5 = x$

2. At what point do the straight lines $2x + 8 = 5y$ and $3y - 5 = x$ intersect?

3. Solve the equations simultaneously: $3x^2 + 5y = y^2 + 18$, $y + 2x = 7$

4. At what point(s) do the straight line $2y - x = 5$ and the curve $x^2 + y^2 = 2y - x$ intersect?

5. Solve the equations for x and y: $x^2 + y^2 = 5$, $x^2 - y^2 = 3$

6. Solve the equations for A, B, and C.

$$A + B = 0$$
$$A + 2C = 5$$
$$3B - 2C = 7$$

7. Find the y-coordinates of the points at which the curves $3y - x = 9$ and $x = 2y^2 - 5y - 3$ intersect.

8. Solve the equations for A, B, C, and D.

$$3A + C = 2$$
$$B - 2C + D = -6$$
$$A - C - 4D = 4$$
$$3A + C - 2D = 6$$

9. Solve the equations for A, B, C, and D.

$$A + D = -1$$
$$A + C + D = 0$$
$$B - D = 2$$
$$2B + C - D = 2$$

10 Completing the Square

EXAMPLE 1

Completing the Square

Write $2x^2 + 7x + 11$ in the form $p[(x+q)^2 + r]$.

SOLUTION

Factor out the coefficient of x^2.

$$2x^2 + 7x + 11 = 2\left[x^2 + \frac{7x}{2} + \frac{11}{2}\right]$$

Add and subtract $\left(\frac{1}{2}\cdot \text{coefficient of } x\right)^2$. The coefficient of x is $\frac{7}{2}$.

$$2x^2 + 7x + 11 = 2\left[x^2 + \frac{7x}{2} + \left(\frac{1}{2}\cdot\frac{7}{2}\right)^2 - \left(\frac{1}{2}\cdot\frac{7}{2}\right)^2 + \frac{11}{2}\right]$$

$$= 2\left[x^2 + \frac{7x}{2} + \left(\frac{7}{4}\right)^2 - \left(\frac{7}{4}\right)^2 + \frac{11}{2}\right]$$

$$= 2\left[\left(x + \frac{7}{4}\right)^2 - \frac{49}{16} + \frac{11}{2}\right]$$

$$= 2\left[\left(x + \frac{7}{4}\right)^2 + \frac{-49 + 88}{16}\right]$$

$$= 2\left[\left(x + \frac{7}{4}\right)^2 + \frac{39}{16}\right]$$

SIMILAR PROBLEM

1. Write $3x^2 + 5x + 14$ in the form $p[(x+q)^2 + r]$.

ANSWER

1. $3\left[\left(x + \frac{5}{6}\right)^2 + \frac{143}{36}\right]$

EXAMPLE 2

Completing the Square

Write $\dfrac{-x^2}{3} + \dfrac{4x}{3} + 4$ in the form $s[(x+t)^2 - u^2]$.

SOLUTION

Factor out the coefficient of x^2.

$$\frac{-x^2}{3} + \frac{4x}{3} + 4 = -\frac{1}{3}[x^2 - 4x - 12]$$

Add and subtract $\left(\frac{1}{2}\cdot \text{coefficient of } x\right)^2$.

$$-\frac{x^2}{3} + \frac{4x}{3} + 4 = -\frac{1}{3}\left[x^2 - 4x + \left(\frac{1}{2}\cdot 4\right)^2 - \left(\frac{1}{2}\cdot 4\right)^2 - 12\right]$$

$$= -\frac{1}{3}[x^2 - 4x + 2^2 - 2^2 - 12]$$

$$= -\frac{1}{3}[(x-2)^2 - 4 - 12]$$

$$= -\frac{1}{3}[(x-2)^2 - 16]$$

$$= -\frac{1}{3}[(x-2)^2 - 4^2]$$

SIMILAR PROBLEM

1. Write $\dfrac{-x^2}{4} + \dfrac{3x}{2} - 2$ in the form $s[(x+t)^2 - u^2]$.

ANSWER

1. $-\dfrac{1}{4}[(x-3)^2 - 1^2]$

EXAMPLE 3

Obtaining the Standard Form for a Circle

Write the equation

$$16x^2 + 16y^2 + 56x - 64y - 31 = 0$$

in the form $(x-h)^2 + (y-k)^2 = r^2$ where h, k, and r are constants. This is the standard form of the equation of a circle.

SOLUTION

Collect the two terms containing x and the two containing y.

$$(16x^2 + 56x + \quad) + (16y^2 - 64y + \quad) = 31$$

Divide by the coefficient of x^2 and y^2.

$$\left(x^2 + \frac{7}{2}x + \quad\right) + (y^2 - 4y + \quad) = \frac{31}{16}$$

Add $\left(\frac{1}{2} \cdot \text{coefficient of } x\right)^2$ and $\left(\frac{1}{2} \cdot \text{coefficient of } y\right)^2$ to both sides of the equation.

$$\left(x^2 + \frac{7}{2}x + \frac{49}{16}\right) + (y^2 - 4y + 4) = \frac{31}{16} + \frac{49}{16} + 4$$

Factor the trinomials in x and in y. They will be perfect squares.

$$\left(x + \frac{7}{4}\right)^2 + (y-2)^2 = \frac{31}{16} + \frac{49}{16} + \frac{64}{16}$$

$$\left(x + \frac{7}{4}\right)^2 + (y-2)^2 = \frac{144}{16}$$

$$\left(x + \frac{7}{4}\right)^2 + (y-2)^2 = 9$$

$$\left(x - \left(-\frac{7}{4}\right)\right)^2 + (y-2)^2 = 3^2$$

SIMILAR PROBLEM

1. Write the equation

$$9x^2 + 12x + 9y^2 - 54y + 49 = 0$$

in the form $(x - h)^2 + (y - k)^2 = r^2$.

ANSWER

1. $\left(x - \left(-\dfrac{2}{3}\right)\right)^2 + (y - 3)^2 = 2^2$

CHAPTER 10 EXERCISES

In Exercises 1–7, complete the square.

1. $2x^2 - 3x + 1$

2. $4x^2 - x - 1$

3. $3 - 3x - 2x^2$

4. $3x^2 - 5x$

5. $3x^2 + x$

6. $-2x^2 - 16x + 18$

7. $-\dfrac{3}{5}x^2 + \dfrac{3}{10}x + \dfrac{3}{10}$

In Exercises 8 and 9, rewrite the equation in the form $(x - h)^2 + (y - k)^2 = r^2$.

8. $16x^2 + 16y^2 - 64y + 8x + 16 = 0$

9. $6x^2 - 6x + 6y^2 + 10y + 5 = 0$

10. Write $2x^2 + y^2 - 2x + 3y - 2\frac{1}{4} = 0$ in the form $\dfrac{(x - h)^2}{a^2} + \dfrac{(y - k)^2}{b^2} = 1$.

11

Exponents and Radicals

REVIEW OF FUNDAMENTALS

FORMULAS USED WITH EXPONENTS AND RADICALS

Operation	*Formulas*
1. A common technique	replace a by a^1.
2. Power zero	$a^0 = 1, \quad a \neq 0$
3. Fractional exponent	$a^{m/n} = \sqrt[n]{a^m} = (\sqrt[n]{a})^m$, so $\sqrt[n]{a} = a^{1/n}$
4. Negative exponent	$a^{-t} = \dfrac{1}{a^t}$
5. Products with the same base	$a^s \cdot a^t = a^{s+t}$
6. Quotients with the same base	$\dfrac{a^s}{a^t} = a^{s-t}$ and $\dfrac{1}{a^{t-s}}$
7. Exponent − exponent	$(a^s)^t = a^{st}$
8. Products to an exponent	$(ab)^s = a^s b^s$, so $\sqrt[n]{ab} = \sqrt[n]{a} \cdot \sqrt[n]{b}$
	$\left[\dfrac{a}{b}\right]^s = \dfrac{a^s}{b^s}$, so $\sqrt[n]{\dfrac{a}{b}} = \dfrac{\sqrt[n]{a}}{\sqrt[n]{b}}$

EXAMPLE 1

Common Errors with Radicals

Are the following statements true or false?

(a) $(\sqrt{x})^2 = x$ for positive x **(b)** $\sqrt{x^2} = x$

(c) $\sqrt{-x}$ is never a real number.   **(d)** $\sqrt{x^2 + 1} = x + 1$

SOLUTION

(a) True. This equation follows from the definition of $\sqrt{x}$: the nonnegative number whose square is x.

(b) False. For example, when $x = -1$, $\sqrt{(-1)^2} = \sqrt{1} = 1 \neq -1$.

(c) False. For example, when $x = -1$, $\sqrt{-(-1)} = \sqrt{1} = 1$ is a real number.

(d) False. There is a temptation to set $\sqrt{x^2 + 1}$ equal to $\sqrt{x^2} + \sqrt{1}$, but this is incorrect. The square root of a sum cannot generally be simplified.

SIMILAR PROBLEM ━━━━━━━━━━━━━━━━━━

1. Is the following statement true or false?

$$\sqrt{x^2 + 4} = x + 2$$

ANSWER

1. False

EXAMPLE 2
━━━━━━━

Rewriting Radicals Using Fractional Exponents

Write the following in the form x^n.

(a) $\sqrt[3]{x}$

(b) $\dfrac{1}{x}$

(c) $\dfrac{1}{\sqrt{x}}$

(d) $\dfrac{\sqrt[3]{x}}{x}$

SOLUTION

(a) $\sqrt[3]{x} = x^{1/3}$

(b) $\dfrac{1}{x} = \dfrac{1}{x^1} = x^{-1}$

(c) $\dfrac{1}{\sqrt{x}} = \dfrac{1}{x^{1/2}} = x^{-1/2}$

(d) $\dfrac{\sqrt[3]{x}}{x} = \dfrac{x^{1/3}}{x} = \dfrac{x^{1/3}}{x^1} = x^{(1/3)-1} = x^{-2/3}$

SIMILAR PROBLEMS ━━━━━━━━━━━━━━━━

Convert the following to the form x^n or u^n.

1. $\sqrt[4]{x}$

2. $\dfrac{1}{x^2}$

3. $\dfrac{1}{\sqrt[3]{x}}$

4. $\dfrac{\sqrt[4]{u}}{u}$

ANSWERS

1. $x^{1/4}$

2. x^{-2}

3. $x^{-1/3}$

4. $u^{-3/4}$

EXAMPLE 3 **Rewriting Radicals Using Fractional Exponents**

Rewrite using fractional exponents.

(a) $\sqrt{1 + x^2}$

(b) $\dfrac{1}{\sqrt{(1 + z^2)^3}}$

SOLUTION

(a) $\sqrt{1 + x^2} = (1 + x^2)^{1/2}$

Note that $1 + x$ is not the answer.

(b) $\dfrac{1}{\sqrt{(1 + z^2)^3}} = \dfrac{1}{[(1 + z^2)^3]^{1/2}}$ Formula 3

$= \dfrac{1}{(1 + z^2)^{3/2}}$ Formula 7

$= (1 + z^2)^{-3/2}$ Formula 4

SIMILAR PROBLEM

Rewrite the following using fractional exponents.

1. $\sqrt{1 - w^2}$

2. $\dfrac{1}{\sqrt[3]{(1 - k^2)^5}}$

ANSWER

1. $(1 - w^2)^{1/2}$

2. $(1 - k^2)^{-5/3}$

EXAMPLE 4 **Evaluating an Expression Involving Exponents**

Evaluate the exponential expression $(x^3 + 17)^{-3/4}$ when $x = 4$.

SOLUTION

$$(x^3 + 17)^{-3/4} = (4^3 + 17)^{-3/4}$$
$$= (64 + 17)^{-3/4}$$
$$= (81)^{-3/4}$$
$$= \dfrac{1}{(81)^{3/4}} \quad \text{Formula 4}$$
$$= \dfrac{1}{(\sqrt[4]{81})^3} \quad \text{Formula 3}$$
$$= \dfrac{1}{3^3}$$
$$= \dfrac{1}{27}$$

SIMILAR PROBLEM

1. Evaluate $(x^2 - 9)^{-2/3}$ when $x = 6$.

ANSWER

1. $\dfrac{1}{9}$

EXAMPLE 5

Simplifying Radicals

Show that $\dfrac{\sqrt{28}}{2} = \sqrt{7}$.

SOLUTION

Factor 28 so that one factor is a perfect square.

$$
\begin{aligned}
\frac{\sqrt{28}}{2} &= \frac{\sqrt{4 \cdot 7}}{2} \qquad \text{Formula 8}\\
&= \frac{\sqrt{4}\sqrt{7}}{2}\\
&= \frac{2 \cdot \sqrt{7}}{2}\\
&= \sqrt{7}
\end{aligned}
$$

SIMILAR PROBLEM

1. Show that $\dfrac{\sqrt{45}}{3} = \sqrt{5}$.

ANSWER

1. $\dfrac{\sqrt{45}}{3} = \dfrac{\sqrt{9 \cdot 5}}{3} = \dfrac{\sqrt{9}\sqrt{5}}{3} = \dfrac{3\sqrt{5}}{3} = \sqrt{5}$

EXAMPLE 6

A Standard Radical Form

Express $\sqrt{4x^2 - 9}$ in the form $k\sqrt{x^2 - a^2}$.

SOLUTION

$$
\begin{aligned}
\sqrt{4x^2 - 9} &= \sqrt{4\left(x^2 - \tfrac{9}{4}\right)} \qquad \text{Factor out the coefficient of } x^2\\
&= \sqrt{4}\sqrt{x^2 - \tfrac{9}{4}} \qquad \text{Formula 8}\\
&= 2\sqrt{x^2 - \left(\tfrac{3}{2}\right)^2}
\end{aligned}
$$

SIMILAR PROBLEM ▬▬▬▬▬▬▬▬▬▬

1. Rewrite $\sqrt{9x^2 - 16}$ in the form $k\sqrt{x^2 - a^2}$.

ANSWER

1. $3\sqrt{x^2 - \left(\frac{4}{3}\right)^2}$

EXAMPLE 7
▬▬▬▬▬▬

Rewriting an Expression Involving Radicals

Rewrite as a sum using fractional exponents.

$$\frac{u^2 - 3u + 2}{\sqrt{u}}$$

SOLUTION

$$
\begin{aligned}
\frac{u^2 - 3u + 2}{\sqrt{u}} &= \frac{u^2 - 3u + 2}{u^{1/2}} \\
&= \frac{u^2}{u^{1/2}} - \frac{3u}{u^{1/2}} + \frac{2}{u^{1/2}} \\
&= \frac{u^2}{u^{1/2}} - \frac{3u^1}{u^{1/2}} + \frac{2}{u^{1/2}} \\
&= u^{2-(1/2)} - 3u^{1-(1/2)} + 2u^{-1/2} \\
&= u^{3/2} - 3u^{1/2} + 2u^{-1/2}
\end{aligned}
$$

▬▬▬

SIMILAR PROBLEM ▬▬▬▬▬▬▬▬▬▬

Rewrite as a sum using fractional exponents.

1. $\dfrac{u^3 + 5u - 3}{\sqrt[3]{u}}$

ANSWER

1. $u^{8/3} + 5u^{2/3} - 3u^{-1/3}$

EXAMPLE 8   **Sketching the Graph of an Exponential Function**

Sketch the graph of $f(x) = x^{2/3}$.

SOLUTION

We begin by plotting several points.

x	-8	-1	0	1	8
y	$(-8)^{2/3} = (\sqrt[3]{-8})^2$ $= (-2)^2$ $= 4$	$(-1)^{2/3} = \sqrt[3]{(-1)^2}$ $= \sqrt[3]{1}$ $= 1$	$0^{2/3} = 0$	$1^{2/3} = 1$	$8^{2/3} = (\sqrt[3]{8})^2$ $= 2^2$ $= 4$

Then we connect the points $(-8, 4)$, $(-1, 1)$, $(0, 0)$, $(1, 1)$, and $(8, 4)$ by a curve as shown in Figure 11.1.

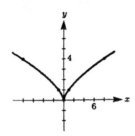

FIGURE 11.1

EXAMPLE 9   **Sketching the Graph of an Exponential Function**

Sketch the graph of $g(x) = 2^x$ and give its domain and range.

SOLUTION

First, we plot a few points.

x	-1	0	1
y	$2^{-1} = \frac{1}{2}$	$2^0 = 1$	$2^1 = 2$

Then we connect the points by a smooth curve, as shown in Figure 11.2.

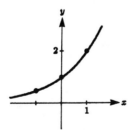

FIGURE 11.2

Since 2^x is always positive, the entire graph is above the x-axis. The domain (the projection of the graph on the x-axis) consists of all real numbers. The range (the projection of the graph on the y-axis) is $(0, \infty)$.

EXAMPLE 10   **The Function** $f(x) = a^x$, $a > 1$

Sketch the graph of $f(x) = a^x$, $a > 1$.

SOLUTION

First we plot a few points.

x	-1	0	1
y	$a^{-1} = \dfrac{1}{a} < 1$	$a^0 = 1$	$a^1 = a > 1$

Then we connect the points by a smooth curve, as shown in Figure 11.3.

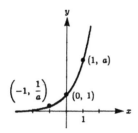

FIGURE 11.3

As in Example 9, the graph is entirely above the x-axis, the domain consists of all real numbers, and the range is $(0, \infty)$.

SIMILAR PROBLEM

1. Sketch the graph of $f(x) = 3^x$.

ANSWER

1.

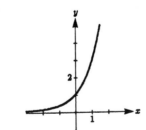

CHAPTER 11 EXERCISES

In Exercises 1–5, rewrite the expression using fractional exponents.

1. $\sqrt[5]{3x}$

2. $\dfrac{\sqrt{x}}{x}$

3. $\dfrac{\sqrt{x}}{\sqrt[3]{x}}$

4. $\dfrac{1}{\sqrt[3]{9+x^2}}$

5. $\dfrac{1}{\sqrt{\sqrt{x}+1}}$

In Exercises 6–10, evaluate the expression at the specified value of x.

6. $(3x^3+3)^{-2/3}$, $x=2$

7. $(4x^2-9)^{-4/3}$, $x=3$

8. $(3x^2-2x-8)^{-3/5}$, $x=4$

9. $\dfrac{1}{(x^2+4)^{-2/3}}$, $x=2$

10. $\dfrac{1}{(\sqrt[3]{x}+6)^{2/3}}$, $x=8$

11. Express $\sqrt{18}$ in terms of $\sqrt{2}$.

12. Show that $\dfrac{\sqrt{54}}{3}=\sqrt{6}$.

13. Show that $\dfrac{\sqrt{48}}{\sqrt{3}}=4$.

14. Express $\dfrac{\sqrt{72}}{6}$ in simplest radical form.

15. Show that $\dfrac{\sqrt{108}}{9}=\dfrac{2}{\sqrt{3}}$.

16. Express $\sqrt{9x^2+36}$ in the form $k\sqrt{x^2+a^2}$.

17. Show that $\sqrt[3]{27x^3-8}=3\sqrt[3]{x^3-\left(\frac{2}{3}\right)^3}$.

18. Express $\sqrt{\dfrac{x^2}{4}+4}$ in the form $k\sqrt{x^2+a^2}$.

19. Show that $\sqrt{1+\dfrac{4}{9x^{2/3}}}=\dfrac{\sqrt{9x^{2/3}+4}}{3x^{1/3}}$.

20. Show that $\sqrt{1+\left(\dfrac{x}{2}-\dfrac{1}{2x}\right)^2}=\dfrac{x}{2}+\dfrac{1}{2x}$.

21. Express $\sqrt{23x^2-5}$ in the form $k\sqrt{x^2-a^2}$.

22. Show that $\sqrt{\left(\dfrac{\sqrt{x}}{2}+\dfrac{1}{2\sqrt{x}}\right)^2-1}=\dfrac{\sqrt{x}}{2}-\dfrac{1}{2\sqrt{x}}$.

In Exercises 23–26, write the expression as a sum of terms.

23. $\dfrac{u+1}{\sqrt[3]{u}}$

24. $\dfrac{2u^4+3u^2-1}{\sqrt{u}}$

25. $\dfrac{u^{1/2}+u^{1/3}+1}{\sqrt{u}}$

26. $\dfrac{5x^3-2\sqrt{x}-4}{\sqrt{x}}$

In Exercises 27–30, sketch the graph of the function.

27. $f(x)=2^{x+1}$

28. $g(x)=3^{-x}$

29. $k(x)=a^{-x}$, $a>0$

30. The inverse function of $h(x)=a^x$, $a>0$

Length, Area, and Volume Formulas

REVIEW OF FUNDAMENTALS

Length, Area, and Volume Formulas

LENGTHS

Pythagorean Theorem	**Similar Triangles**	**Circumference of Circle**
$c^2 = a^2 + b^2$	$\dfrac{a}{b} = \dfrac{a'}{b'}$	$C = 2\pi r$

 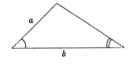

AREAS

Area of Circle

$A = \pi r^2$

Surface Area of Sphere

$A = 4\pi r^2$

Area of Triangle

$A = \dfrac{1}{2}bh$

Area of Trapezoid

$A = \dfrac{1}{2}h(a + b)$

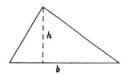 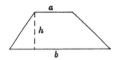

VOLUMES

Volume of Cylinder

$$V = \pi r^2 h$$

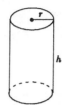

Volume of Sphere

$$V = \frac{4}{3}\pi r^3$$

Volume of Cone

$$V = \frac{1}{3}\pi r^2 h$$

EXAMPLE 1

Finding the Surface Area of a Box

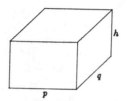

FIGURE 12.1

Find the surface area of a box of height h whose base dimensions are p and q, and that satisfies either one of the following conditions.

(a) The box is closed. (See Figure 12.1.)

(b) The box has an open top.

(c) The box has an open top and a square base.

SOLUTION

(a) The surface area of the box consists of the areas of its six sides, as follows.

$$\begin{aligned}
\text{Surface Area} &= \left(\begin{array}{c}\text{Area}\\ \#1\end{array}\right) + \left(\begin{array}{c}\text{Area}\\ \#2\end{array}\right) + \left(\begin{array}{c}\text{Area}\\ \#3\end{array}\right) + \left(\begin{array}{c}\text{Area}\\ \#4\end{array}\right) + \left(\begin{array}{c}\text{Area}\\ \#5\end{array}\right) + \left(\begin{array}{c}\text{Area}\\ \#6\end{array}\right)\\
&= hp \quad + \quad hp \quad + \quad hq \quad + \quad hq \quad + \quad pq \quad + \quad pq\\
&= 2hp \quad + \quad 2hq \quad + \quad 2pq
\end{aligned}$$

(b) If the box has an open top, then surface #5 is missing, and the surface area is as follows.

$$\begin{aligned}
\text{Surface Area} &= \left(\begin{array}{c}\text{Area}\\ \#1\end{array}\right) + \left(\begin{array}{c}\text{Area}\\ \#2\end{array}\right) + \left(\begin{array}{c}\text{Area}\\ \#3\end{array}\right) + \left(\begin{array}{c}\text{Area}\\ \#4\end{array}\right) + \left(\begin{array}{c}\text{Area}\\ \#6\end{array}\right)\\
&= hp \quad + \quad hp \quad + \quad hq \quad + \quad hq \quad + \quad pq\\
&= 2hp \quad + \quad 2hq \quad + \quad pq
\end{aligned}$$

(c) If the box also has a square base, then we can obtain the surface area by replacing q by p in the formula given in part (b).

$$\text{Surface Area} = 2hp + 2hp + p^2 = 4hp + p^2$$

SIMILAR PROBLEM ▬▬▬▬▬▬▬▬▬

1. A box with height h has a square base of width x. It is missing two sides (not the base or top). Find its surface area.

ANSWER

1. $2x^2 + 2xh$

EXAMPLE 2
▬▬▬▬▬

Finding the Surface Area of a Cylinder

Find the surface area of a cylinder of radius r and height h and which satisfies either one of the following conditions.

(a) The cylinder is closed.

(b) The cylinder has an open top.

SOLUTION

(a) To find the surface area, imagine cutting the (hollow) cylinder into three pieces. The top and base are cut out to make two circles, as shown in Figure 12.2. Adding the areas of these circles we obtain $\pi r^2 + \pi r^2 = 2\pi r^2$. To find the surface area of the side of the cylinder, we imagine that it is cut and rolled out flat to form a rectangle whose length is $2\pi r$ (the circumference) and whose height is h. The area of the rectangle is $2\pi rh$. Thus, the total surface area of the cylinder is

$$\text{Surface Area } = 2\pi r^2 + 2\pi rh.$$

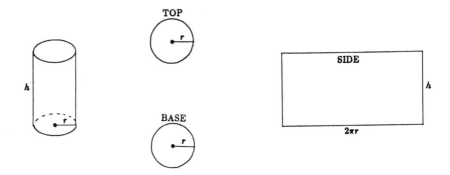

FIGURE 12.2

(b) If the cylinder has an open top, then we must add only the area of one of the circles to the area of the side, obtaining

$$\text{Surface Area } = \pi r^2 + 2\pi rh.$$

▬▬▬▬

SIMILAR PROBLEM ■■■■■■■■■■

1. Find the surface area of the cylinder of radius r and height h which has neither a top nor a base.

ANSWER

1. $2\pi rh$

EXAMPLE 3

Using the Pythagorean Theorem

A seven-foot ladder, leaning against a wall, touches the wall x feet from the ground. Obtain an expression for the distance along the ground from the wall to the ladder.

SOLUTION

Let the distance along the ground be y. Applying the Pythagorean theorem, we obtain

$$x^2 + y^2 = 7^2$$
$$y^2 = 49 - x^2$$
$$y = \sqrt{49 - x^2}.$$

See Figure 12.3.

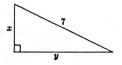

FIGURE 12.3   ■■■■

SIMILAR PROBLEM ■■■■■■■■■■

1. A five-foot ladder, leaning against a wall, has one end t feet from the wall. Find an expression for the height of the other end.

ANSWER

1. $\sqrt{25 - t^2}$

EXAMPLE 4   **Using Similar Triangles**

A 6′ person is standing x feet away from a 10′ lamppost. What is the distance d from the base of the lamppost to the end of the person's shadow, expressed as a function of x?

SOLUTION

A pair of triangles is similar if two angles of one triangle are each equal to two of the other. The situation is illustrated in Figure 12.4. Triangles ABE and ACD are similar because they both contain angle A and they both contain a right angle.

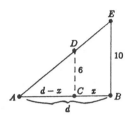

FIGURE 12.4

We solve the problem using the following basic property of similar triangles.

$$\frac{AC}{6} = \frac{AB}{10}$$

$$\frac{d - x}{6} = \frac{d}{10}$$

$$\frac{\text{Side whose length is } AC}{\text{Side whose length is } 6} = \frac{\text{Side whose length is } AB}{\text{Side whose length is } 10}$$

Now we solve for d.

$$\frac{d - x}{6} = \frac{d}{10}$$

$$5d - 5x = 3d$$

$$2d = 5x$$

$$d = \frac{5x}{2}$$

SIMILAR PROBLEM

1. A 6′ person is standing x feet away from an 11′ lamppost. What is the distance d from the base of the lamppost to the end of the person's shadow, expressed as a function of x?

ANSWER

1. $d = \dfrac{11x}{5}$

EXAMPLE 5   **Finding an Expression for Area**

A piece of wire 5 inches long is to be cut into two pieces. One piece is x inches long and is to be bent into the shape of a square. The other piece is to be bent into the shape of a circle. Find an expression for the total area made up by the square and the circle as a function of x.

SOLUTION

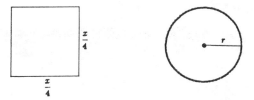

FIGURE 12.5

Since the length of the first piece is x, the length of the second piece must be $5 - x$. First we find the area of the square. The four equal sides of the square must add up to x, so each must have length $x/4$. (See Figure 12.5.) The area of the square is

$$\text{Area} = \left(\frac{x}{4}\right)\left(\frac{x}{4}\right) = \frac{x^2}{16}.$$

Next we find the area of the circle. To find the area of the circle, we must find its radius r. We know that the circumference must equal $5 - x$, so we can say

$$2\pi r = 5 - x$$
$$r = \frac{5 - x}{2\pi}.$$

Therefore, the area of the circle is

$$\text{Area} = \pi r^2$$
$$= \pi\left(\frac{5 - x}{2\pi}\right)^2$$
$$= \frac{\pi(5 - x)^2}{4\pi^2}$$
$$= \frac{5^2 - 2\cdot 5x + x^2}{4\pi}.$$

Finally, adding the two areas, we find the total area to be

$$\text{Total area} = \frac{x^2}{16} + \frac{25 - 10x + x^2}{4\pi}.$$

SIMILAR PROBLEM

1. A piece of wire 7 inches long is to be cut into two pieces. One piece is t inches long and is to be bent into the shape of a circle. The other piece is to be bent into the shape of a square. Find an expression for the total area made up by the square and the circle.

ANSWER

1. $\dfrac{t^2}{4\pi} + \dfrac{49 - 14t + t^2}{16}$

CHAPTER 12 EXERCISES

1. Find an expression for the printed area of a page with the following specifications. The page has a width of w'' and a height of h''. The margins on the sides are 1 inch each, and the margins at the top and bottom are $\frac{3}{2}$ inches each.

2. Find an expression for the area of the window as shown. It consists of a rectangle of height h'' and width w'', with a semi-circle mounted on top of the rectangle.

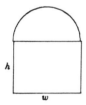

FIGURE FOR 2

3. A ship is anchored five miles from the nearest point P on shore. A person plans to row in a straight line from the ship to a point x miles downstream from P. Find an expression for the distance the person will have to row.

4. We are given a rectangular piece of cardboard, measuring 7″ by 9″. Four identical squares of length x are cut from the four corners, as shown, and the resulting piece of cardboard is folded along the dotted lines shown to make a box. Express the volume of the box as a function of x.

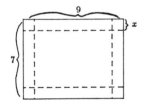

FIGURE FOR 4

5. A cylinder of height h is inscribed in a sphere of radius q. (In other words, the edges of the top and base of the cylinder are touching the sphere.) Find an expression for the volume of the cylinder.

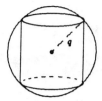

FIGURE FOR 5

6. A nine-inch piece of wire is cut into two pieces. One of the pieces, which has a length t, is bent into a square and the other is bent into an equilateral triangle. Find an expression for the total area made up of the square and triangle.

7. An eight-inch piece of wire is cut into two pieces. One of the pieces has length t and is bent into the shape of an equilateral triangle. The other is bent into the shape of a circle. Express the total area A obtained as a function of t.

8. A 6′ person is standing x feet away from a 20′ lamppost. What is the distance t from the man's feet to the end of his shadow, expressed as a function of x?

9. A garage light 9′ high is 5′ from the door. Suppose that the base of the partially open door is r feet from the ground, and that the door's shadow is s feet from the garage. Express s as a function of r.

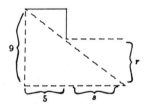

FIGURE FOR 9

10. A ladder is leaning against a wall and touches the top of a 7′ fence which is 2′ away from the wall. Express y, the length of the ladder required, as a function of x, the distance along the ground from the ladder to the fence.

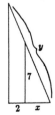

FIGURE FOR 10

13 Logarithms

REVIEW OF FUNDAMENTALS

■ The number $\log_b x$ is the **exponent** to which b must be raised to give x. Formally, $\log_b x$ is defined by $b^{\log_b x} = x$. The following rules can be derived from this definition.

■ **Logarithm Formulas**

1. $\log_b b = 1$ because $b^1 = b$.

2. $\log_b 1 = 0$ because $b^0 = 1$.

3. **Products:** $\log_b cd = \log_b c + \log_b d$

4. **Quotients:** $\log_b \dfrac{c}{d} = \log_b c - \log_b d$

5. **Powers:** $\log_b c^t = t \cdot \log_b c$

6. **Change of Base:** $\log_k x = \log_k b \cdot \log_b x = \dfrac{\log_b x}{\log_b k}$

EXAMPLE 1

Evaluating a Logarithm

Evaluate $\log_9 3$.

SOLUTION

$\log_9 3$ is the *exponent* to which 9 must be raised to give 3.

$$9^x = 3$$

Since $9^{1/2} = \sqrt{9} = 3$, the answer is

$$\log_9 3 = \frac{1}{2}.$$

SIMILAR PROBLEM

1. Evaluate $\log_8 2$.

ANSWER

1. $\dfrac{1}{3}$

EXAMPLE 2

Simplifying a Logarithm

Simplify the expression $\log_b b^{5x+1}$.

SOLUTION

$\log_b b^{5x+1}$ is the *exponent* to which b must be raised to give b^{5x+1}. Thus, the answer is $5x + 1$. That is,

$$\log_b b^{5x+1} = 5x + 1.$$

SIMILAR PROBLEM

1. Simplify the expression $\log_b b^{7x-2}$.

ANSWER

1. $7x - 2$

EXAMPLE 3

Simplifying a Logarithmic Expression

Simplify the expression $a^{\log_a x}$.

SOLUTION

$\log_a x$ is the exponent to which a must be raised to give x. Thus,

$$a^{\log_a x} = x.$$

SIMILAR PROBLEM

1. Simplify the expression $k^{\log_k w}$.

ANSWER

1. w

EXAMPLE 4

The Function $f(x) = \log_b x, \quad b > 1$

Sketch the graph of $f(x) = \log_b x$, and give its domain and range.

SOLUTION

The statement $y = \log_b x$ is equivalent to $x = b^y$. Substitute y for x and we have $y = b^x$. The function $f(x) = \log_b x$ is the inverse of the function $f^{-1}(x) = b^x$. Therefore, to sketch the graph of $y = \log_b x$, we can first sketch the graph of $y = b^x$. Then, reflect the graph about the line $y = x$, as shown in Figure 13.1. The domain (projection of the graph onto the x-axis) consists of all positive real numbers and the range consists of all real numbers.

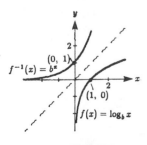

FIGURE 13.1

EXAMPLE 5

Changing the Base of the Log

Using the fact that $\log_{10} 2 \approx 0.3010$, express $\log_{10}(x^2+1)$ in terms of $\log_2(x^2+1)$.

SOLUTION

Applying Formula 6 with $b = 2$ and $k = 10$, we have

$$\log_k x = \log_k b \cdot \log_b x$$
$$\log_{10}(x^2 + 1) = \log_{10} 2 \cdot \log_2(x^2 + 1)$$
$$\approx 0.3010 \log_2(x^2 + 1).$$

SIMILAR PROBLEM

1. Using the fact that $\log_{10} 3 \approx 0.47$, express $\log_{10}(x^2 - 1)$ in terms of $\log_3(x^2 - 1)$.

ANSWER

1. $0.47 \log_3(x^2 - 1)$

EXAMPLE 6

Expanding a Logarithmic Expression

Write the expression as a sum of logarithms.

$$\log_b \frac{(4x^5 - x - 1)\sqrt{x - 7}}{\sqrt[3]{x^2 + 1}}$$

SOLUTION

$$\log_b \frac{(4x^5 - x - 1)\sqrt{x - 7}}{\sqrt[3]{x^2 + 1}} = \log_b[(4x^5 - x - 1)\sqrt{x - 7}] - \log_b \sqrt[3]{x^2 + 1}$$
$$= \log_b(4x^5 - x - 1) + \log_b \sqrt{x - 7} - \log_b(x^2 + 1)^{1/3}$$
$$= \log_b(4x^5 - x - 1) + \frac{1}{2}\log_b(x - 7) - \frac{1}{3}\log_b(x^2 + 1)$$

SIMILAR PROBLEM

1. Write the following as a sum of logarithms.

$$\log_b \frac{(3x^6 + 2)\sqrt{x + 8}}{\sqrt[4]{x - 1}}$$

ANSWER

1. $\log_b(3x^6 + 2) + \dfrac{1}{2}\log_b(x + 8) - \dfrac{1}{4}\log_b(x - 1)$

EXAMPLE 7   **Condensing a Logarithmic Expression**

Write the expression as a logarithm of a single expression.

$$\log_a(3x+2) - \frac{1}{3}\log_a(2x+1) - 7\log_a(x^4+x+1)$$

SOLUTION

$$\log_a(3x+2) - \frac{1}{3}\log_a(2x+1) - 7\log_a(x^4+x+1)$$
$$= \log_a(3x+2) - \log_a(2x+1)^{1/3} - \log_a(x^4+x+1)^7$$
$$= \log_a(3x+2) - [\log_a(2x+1)^{1/3} + \log_a(x^4+x+1)^7]$$
$$= \log_a(3x+2) - \log_a[(2x+1)^{1/3}(x^4+x+1)^7]$$
$$= \log_a \frac{(3x+2)}{(2x+1)^{1/3}(x^4+x+1)^7}$$
$$= \log_a \frac{(3x+2)}{\sqrt[3]{2x+1}(x^4+x+1)^7}$$

SIMILAR PROBLEM

1.   Write the expression as a logarithm of a single expression.

$$\log_a(2x-1) - \frac{1}{2}\log_a(3x-1) - 9\log_a(x^3-8)$$

ANSWER

1. $\log_a \dfrac{(2x-1)}{\sqrt{3x-1}(x^3-8)^9}$

CHAPTER 13 EXERCISES

1. Evaluate $\log_b 1$.

2. Evaluate $\log_k \sqrt{k}$.

3. Suppose that $\log_{10} 5 \approx 0.6990$. Express $\log_{10} x$ in terms of $\log_5 x$.

4. Write $\log_{10}(2x + 1)$ in terms of $\log_p(2x + 1)$.

5. Compute $\log_8 64$.

6. Compute $\log_5(1/25)$.

7. Express $\log_b \dfrac{(2x^2 - 3)(x^3 + 1)^2}{\sqrt{7x + 1}}$ as a sum of logarithms.

8. Write as a logarithm of a single expression.

$$\frac{1}{2}\log_b(3x + 1) - \frac{2}{3}\log_b(1 - 9x^2) + \log_b(1 - x)$$

9. Write as a logarithm of a single expression.

$$\frac{3}{2}\log_b(2 + x)^2 - \frac{5}{2}\log_b(x^2 - 4) - \frac{2}{3}\log_b 27x + \frac{1}{3}\log_b 8x^2$$

10. Sketch the graph of $f(x) = \log_2 x$.

11. Sketch the graph of $g(x) = \log_3 x$.

12. By choosing appropriate values of a, b, and c, show that $\log_b(a + b) = \log_b a + \log_b c$ is false.

14

Trigonometry I: Trigonometric Functions and Their Inverses

REVIEW OF FUNDAMENTALS

■ **Definitions of Trigonometric Ratios for Acute Angles**

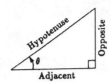

$$\sin \theta = \frac{\text{Opposite}}{\text{Hypotenuse}} \qquad \cos \theta = \frac{\text{Adjacent}}{\text{Hypotenuse}} \qquad \tan \theta = \frac{\text{Opposite}}{\text{Adjacent}}$$

Memory aid: "SOH-CAH-TOA"

$$\csc \theta = \frac{\text{Hypotenuse}}{\text{Opposite}} \qquad \sec \theta = \frac{\text{Hypotenuse}}{\text{Adjacent}} \qquad \cot \theta = \frac{\text{Adjacent}}{\text{Opposite}}$$

■ **Radian-Degree Conversions**

$180° = \pi$ radians

■ **Trigonometric Functions of Common Angles**

$\theta = \dfrac{\pi}{3}$ or $\theta = \dfrac{\pi}{6}$

$$\sin \frac{\pi}{3} = \frac{\sqrt{3}}{2} \qquad\qquad \cos \frac{\pi}{3} = \frac{1}{2}$$

$$\sin \frac{\pi}{6} = \frac{1}{2} \qquad\qquad \cos \frac{\pi}{6} = \frac{\sqrt{3}}{2}$$

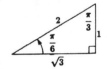

$\theta = \dfrac{\pi}{4}$

$$\sin \frac{\pi}{4} = \frac{\sqrt{2}}{2} \qquad\qquad \cos \frac{\pi}{4} = \frac{\sqrt{2}}{2}$$

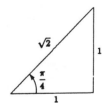

■ Finding the Trigonometric Ratio of Any Angle

1. Sketch the angle with its vertex at the origin, and its initial side (at which the rotation begins) on the positive x-axis. The formal definitions are: $\sin \theta = y/r$, $\cos \theta = x/r$ and $\tan \theta = y/x$ where (x, y) is any point on the terminal side, and $r = \sqrt{x^2 + y^2}$.

2. Compute the reference angle θ' (the **acute** angle made by the terminal side and the x-axis), as shown in Figure 14.1.

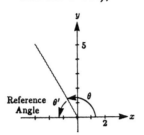

FIGURE 14.1

3. Use the reference angle θ' to find the value of a trigonometric function of θ.

    Terminal side in Quadrant II:

    $\sin \theta = \sin \theta'$ $\qquad$ $\cos \theta = - \cos \theta'$ $\qquad$ $\tan \theta = - \tan \theta'$

    Terminal side in Quadrant III:

    $\sin \theta = - \sin \theta'$ $\qquad$ $\cos \theta = - \cos \theta'$ $\qquad$ $\tan \theta = \tan \theta'$

    Terminal side in Quadrant IV:

    $\sin \theta = - \sin \theta'$ $\qquad$ $\cos \theta = \cos \theta'$ $\qquad$ $\tan \theta = - \tan \theta'$

■ Graphs of Trigonometric Functions

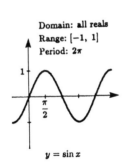

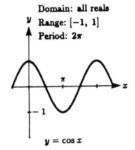

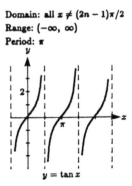

■ The Inverse Trigonometric Functions

Function	Domain	Range
$y = \arcsin x$ iff $\sin y = x$	$-1 \le x \le 1$	$-\dfrac{\pi}{2} \le y \le \dfrac{\pi}{2}$
$y = \arccos x$ iff $\cos y = x$	$-1 \le x \le 1$	$0 \le y \le \pi$
$y = \arctan x$ iff $\tan y = x$	$-\infty < x < \infty$	$-\dfrac{\pi}{2} < y < \dfrac{\pi}{2}$
$y = \text{arccot } x$ iff $\cot y = x$	$-\infty < x < \infty$	$0 < y < \pi$
$y = \text{arcsec } x$ iff $\sec y = x$	$x \le -1, \ 1 \le x$	$0 \le y \le \pi, \ y \neq \dfrac{\pi}{2}$
$y = \text{arccsc } x$ iff $\csc y = x$	$x \le -1, \ 1 \le x$	$-\dfrac{\pi}{2} \le y \le \dfrac{\pi}{2}, \ y \neq 0$

EXAMPLE 1

Common Values of Trigonometric and Inverse Trigonometric Functions

Evaluate the following.

(a) $\cos 0$ **(b)** $\sin 0$ **(c)** $\tan \dfrac{\pi}{2}$ **(d)** $\cos \dfrac{\pi}{4}$

(e) $\csc \dfrac{\pi}{3}$ **(f)** $\arccos \dfrac{\sqrt{3}}{2}$ **(g)** $\arctan 1$

SOLUTION

(a) $\cos 0 = 1$

(b) $\sin 0 = 0$

(c) $\tan \dfrac{\pi}{2}$ is undefined.

(d) $\cos \dfrac{\pi}{4} = \dfrac{\sqrt{2}}{2}$

(e) $\csc \dfrac{\pi}{3} = \dfrac{1}{\sin(\pi/3)} = \dfrac{1}{\sqrt{3}/2} = \dfrac{2}{\sqrt{3}}$

(f) $\quad y = \arccos \dfrac{\sqrt{3}}{2}$

$\qquad \cos y = \dfrac{\sqrt{3}}{2}, \quad 0 \le y \le \pi$

$\qquad y = \dfrac{\pi}{6}$

(g) $\quad y = \arctan 1$

$\qquad \tan y = 1, \quad -\dfrac{\pi}{2} < y < \dfrac{\pi}{2}$

$\qquad y = \dfrac{\pi}{4}$

SIMILAR PROBLEMS

Evaluate the following.

1. $\sin \dfrac{\pi}{2}$ **2.** $\cos \dfrac{\pi}{2}$ **3.** $\tan\left(-\dfrac{\pi}{2}\right)$ **4.** $\sin \dfrac{\pi}{4}$

5. $\sec \dfrac{\pi}{3}$ **6.** $\arcsin \dfrac{\sqrt{3}}{2}$ **7.** $\arctan \dfrac{1}{\sqrt{3}}$

ANSWERS

1. 1   **2.** 0   **3.** does not exist   **4.** $\dfrac{\sqrt{2}}{2} = \dfrac{1}{\sqrt{2}}$

5. 2   **6.** $\dfrac{\pi}{3}$ **7.** $\dfrac{\pi}{6}$

EXAMPLE 2

Trigonometric Ratios of All Angles

Evaluate the following.

(a) $\cos\left(-\dfrac{\pi}{4}\right)$ **(b)** $\sin\dfrac{4\pi}{3}$

(c) $\cos 150°$ **(d)** $\tan\dfrac{2\pi}{3}$

SOLUTION

(a) Reference angle for $\theta = -\pi/4$ is $\theta' = \pi/4$ and the terminal side lies in Quadrant IV.

$$\cos\left(-\frac{\pi}{4}\right) = \cos\frac{\pi}{4} = \frac{\sqrt{2}}{2}$$

(b) Reference angle for $\theta = 4\pi/3$ is $\theta' = \pi/3$ and the terminal side lies in Quadrant III.

$$\sin\frac{4\pi}{3} = -\sin\frac{\pi}{3} = -\frac{\sqrt{3}}{2}$$

(c) Reference angle for $\theta = 150°$ is $\theta' = 30°$ and the terminal side lies in Quadrant II.

$$\cos 150° = -\cos 30° = -\frac{\sqrt{3}}{2}$$

(d) Reference angle for $\theta = 2\pi/3$ is $\theta' = \pi/3$ and the terminal side lies in Quadrant II.

$$\tan\frac{2\pi}{3} = -\tan\frac{\pi}{3} = -\sqrt{3}$$

SIMILAR PROBLEMS

Evaluate the following.

1. $\sin\left(-\dfrac{\pi}{4}\right)$ **2.** $\cos\dfrac{4\pi}{3}$ **3.** $\sin 150°$ **4.** $\cos\dfrac{2\pi}{3}$

ANSWERS

1. $-\dfrac{1}{\sqrt{2}}$ **2.** $-\dfrac{1}{2}$ **3.** $\dfrac{1}{2}$ **4.** $-\dfrac{1}{2}$

EXAMPLE 3

Solving a Trigonometric Equation

For what values of θ is $\cos 4\theta = 0$?

SOLUTION

From the graph of $\cos x$ we observe that $\cos x = 0$ when

$$x = \frac{\pi}{2}, \frac{-\pi}{2}, \frac{3\pi}{2}, \frac{-3\pi}{2}, \ldots \quad \text{or} \quad x = \frac{\pi}{2} + n\pi, \quad n = \pm 1, \pm 2, \ldots .$$

Thus, $\cos 4\theta = 0$ when

$$4\theta = \frac{\pi}{2}, \frac{-\pi}{2}, \frac{3\pi}{2}, \frac{-3\pi}{2}, \ldots \quad \text{or} \quad 4\theta = \frac{\pi}{2} + n\pi.$$

Dividing both sides by 4, we obtain

$$\theta = \frac{\pi}{2} \cdot \frac{1}{4}, \frac{-\pi}{2} \cdot \frac{1}{4}, \frac{3\pi}{2} \cdot \frac{1}{4}, \frac{-3\pi}{2} \cdot \frac{1}{4} \ldots$$
$$= \frac{\pi}{8}, \frac{-\pi}{8}, \frac{3\pi}{8}, \frac{-3\pi}{8}, \ldots \quad \text{or} \quad \theta = \frac{\pi}{8} + \frac{n\pi}{4}.$$

SIMILAR PROBLEM

1. For what values of θ is $\sin 6\theta = 0$?

ANSWER

1. $\theta = \frac{n\pi}{6}$, $n = 0, \pm 1, \pm 2, \ldots$

EXAMPLE 4

Compositions Involving Trigonometric Functions

If $f(x) = 2x - 1$ and $g(x) = \cos x$, find $(g \circ f)(x)$.

SOLUTION

$$(g \circ f)(x) = g(f(x))$$
$$= g(2x - 1)$$
$$= \cos(2x - 1)$$

EXAMPLE 5

Compositions Involving Trigonometric Functions

Express $h(x) = \sin(3x^2 + 5)$ as the composition of two simpler functions.

SOLUTION

The function $h(x) = \sin(3x^2 + 5)$ consists of the sine function applied to the function $3x^2 + 5$. Thus $h = p \circ g$ where $p(x) = \sin x$ and $g(x) = 3x^2 + 5$.

Check:
$$(p \circ g)(x) = p(g(x))$$
$$= p(3x^2 + 5)$$
$$= \sin(3x^2 + 5)$$

SIMILAR PROBLEMS ▬▬▬▬▬▬▬▬▬▬▬▬▬▬▬▬▬

1. If $f(x) = 3x + 1$ and $g(x) = \sin x$, find $(g \circ f)(x)$.

2. Express $h(x) = \cos(x + 1)$ as a composition of two simpler functions.

ANSWERS

1. $(g \circ f)(x) = \sin(3x + 1)$

2. $h = p \circ q$ where $p(x) = \cos x$ and $q(x) = x + 1$.

EXAMPLE 6
▬▬▬▬▬

Understanding Trigonometric Exponent Notation

Which of the expressions are identical?

(a) $\cos^2 x$ (b) $(\cos x)^2$

(c) $\cos(x^2)$ (d) $(\sin x)^{-1}$

(e) $\sin^{-1} x$ (meaning arcsin x)   (f) $\sin(x^{-1})$

(g) $\dfrac{1}{\sin x}$

SOLUTION

There are *only* two pairs that are identical, (a) and (b), and (d) and (g) because of the following definitions.

$$\cos^2 x = (\cos x)^2$$

$$(\sin x)^{-1} = \frac{1}{\sin x}$$

▬▬▬▬▬

SIMILAR PROBLEMS ▬▬▬▬▬▬▬▬▬▬▬▬▬▬▬▬▬

1. Which of the expressions are identical?

(a) $\sin^2 x$ (b) $(\sin x)^2$

(c) $\sin(x^2)$ (d) $(\cos x)^{-1}$

(e) $\cos^{-1} x$ (meaning arccos x)   (f) $\cos(x^{-1})$

(g) $\dfrac{1}{\cos x}$

ANSWERS

1. (a) and (b) are identical, (d) and (g) are identical.

EXAMPLE 7

Trigonometric Functions and Inverse Trigonometric Functions

Express $\sin(\arccos x)$ in a form that contains no trigonometric functions or their inverses.

SOLUTION

To use $\arccos x$, we let $y = \arccos x$. Then, using the definition of $\arccos x$, we have

$$\cos y = x, \quad 0 \le y \le \pi.$$

By writing

$$\cos y = \frac{x}{1} = \frac{\text{adjacent}}{\text{hypotenuse}}$$

we sketch a triangle containing an angle with y radians, an adjacent side of length x, and a hypotenuse of length 1. (See Figure 14.2.)

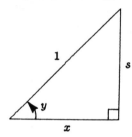

FIGURE 14.2

Next, we find the length of the opposite side, s. From the Pythagorean Theorem, we have

$$x^2 + s^2 = 1^2$$
$$s^2 = 1 - x^2$$
$$s = \sqrt{1 - x^2}.$$

Now, we have

$$\sin(\arccos x) = \sin y$$
$$= \frac{\text{opposite}}{\text{hypotenuse}}$$
$$= \frac{s}{1}$$
$$= \sqrt{1 - x^2}.$$

SIMILAR PROBLEM ▬▬▬▬▬▬▬▬▬▬▬▬▬▬▬

1. Express $\cos(\arcsin x)$ in a form that contains no trigonometric functions or their inverses.

ANSWER

1. $\sqrt{1 - x^2}$

CHAPTER 14 EXERCISES

1. Compute the following.

    (a) $\cos \dfrac{\pi}{3}$

    (b) $\csc \dfrac{\pi}{4}$

    (c) $\tan \dfrac{\pi}{6}$

2. Compute the following.

    (a) $\sin\left(-\dfrac{\pi}{3}\right)$

    (b) $\cos\left(-\dfrac{\pi}{6}\right)$

    (c) $\cos\left(\dfrac{\pi}{2} + \dfrac{\pi}{3}\right)$

    (d) $\sin \dfrac{5\pi}{6}$

    (e) $\cos\left(-\dfrac{3\pi}{4}\right)$

3. Compute the following.

    (a) $\cos^2 \dfrac{\pi}{4}$

    (b) $\left(\cos \dfrac{\pi}{4}\right)^2$

    (c) $\sin^2\left(-\dfrac{\pi}{4}\right)$

    (d) $\tan^2 0$

4. Compute the following.

    (a) $\arcsin \dfrac{\sqrt{3}}{2}$

    (b) $\arccos(-1)$

5. Find the values of θ for which $\cos 6\theta = 0$.

6. Express each of the following functions as the composition of simpler functions.

    (a) $\sin(2x - 1)$ (b) $\arccos(x^2 - 1)$

7. Express $\tan(\arccos(x + 1))$ in a form without trigonometric functions or their inverses.

8. Find the domain and range of $f(x) = \sin x$.

9. Find the domain and range of $g(x) = \tan x$.

15

Trigonometry II: Trigonometric Identities

REVIEW OF FUNDAMENTALS

- **Identities Most Commonly Used in Calculus**

$$\csc x = \frac{1}{\sin x} \quad \sec x = \frac{1}{\cos x} \quad \cot x = \frac{1}{\tan x} \quad \tan x = \frac{\sin x}{\cos x} \quad \cot x = \frac{\cos x}{\sin x}$$

$$\sin^2 x + \cos^2 x = 1 \qquad \tan^2 x + 1 = \sec^2 \qquad 1 + \cot^2 x = \csc^2 x$$

$$\sin^2 x = \frac{1}{2}(1 - \cos 2x) \qquad \cos^2 x = \frac{1}{2}(1 + \cos 2x)$$

$$\sin 2x = 2 \sin x \cos x$$

- **Other Identities**

$$\cos(A - B) = \cos A \cos B + \sin A \sin B$$
$$\cos(A + B) = \cos A \cos B - \sin A \sin B$$

$$\sin(A - B) = \sin A \cos B - \cos A \sin B$$
$$\sin(A + B) = \sin A \cos B + \cos A \sin B$$

$$\sin A \sin B = \frac{1}{2}[\cos(A - B) - \cos(A + B)]$$

$$\cos A \cos B = \frac{1}{2}[\cos(A - B) + \cos(A + B)]$$

$$\sin A \cos B = \frac{1}{2}[\sin(A + B) + \sin(A - B)]$$

$$\cos 2x = \cos^2 x - \sin^2 x = 1 - 2 \sin^2 x = 2 \cos^2 x - 1$$

EXAMPLE 1

Verifying a Trigonometric Identity

Use the trigonometric identity $\sin^2 x + \cos^2 x = 1$ to verify that $\tan^2 x + 1 = \sec^2 x$.

SOLUTION

$$\sin^2 x + \cos^2 x = 1$$

$$\frac{\sin^2 x}{\cos^2 x} + \frac{\cos^2 x}{\cos^2 x} = \frac{1}{\cos^2 x} \qquad \text{Divide both sides by } \cos^2 x$$

$$\left(\frac{\sin x}{\cos x}\right)^2 + 1 = \left(\frac{1}{\cos x}\right)^2$$

$$(\tan x)^2 + 1 = (\sec x)^2$$

$$\tan^2 x + 1 = \sec^2 x$$

SIMILAR PROBLEM

1. Show how to obtain $1 + \cot^2 = \csc^2 x$ from $\sin^2 x + \cos^2 x = 1$.

ANSWER

1. $\sin^2 x + \cos^2 x = 1$

$$1 + \frac{\cos^2 x}{\sin^2 x} = \frac{1}{\sin^2 x}$$

$$1 + \cot^2 x = \csc^2 x$$

EXAMPLE 2

Verifying a Trigonometric Identity

Use the identities

$$\cos(A - B) = \cos A \cos B + \sin A \sin B$$
$$\cos(A + B) = \cos A \cos B - \sin A \sin B$$

to verify that

$$\frac{1}{2}[\cos(A - B) + \cos(A + B)] = \cos A \cos B.$$

SOLUTION

By adding these equations, we obtain

$$\cos(A - B) + \cos(A + B) = 2 \cos A \cos B$$

$$\frac{1}{2}[\cos(A - B) + \cos(A + B)] = \cos A \cos B. \qquad \text{Divide by 2}$$

SIMILAR PROBLEM

1. Use the identities

$$\cos(A - B) = \cos A \cos B + \sin A \sin B$$
$$\cos(A + B) = \cos A \cos B - \sin A \sin B$$

to verify that

$$\frac{1}{2}[\cos(A - B) - \cos(A + B)] = \sin A \sin B.$$

ANSWER

1. Subtract the given equations to obtain

$$\cos(A - B) - \cos(A + B) = 2 \sin A \sin B$$

$$\frac{1}{2}[\cos(A - B) - \cos(A + B)] = \sin A \sin B. \qquad \text{Divide by 2}$$

EXAMPLE 3

Verifying a Trigonometric Identity

Use the trigonometric identity

$$\cos(A + B) = \cos A \cos B - \sin A \sin B$$

to verify that

$$\cos 2x = \cos^2 - \sin^2 x.$$

SOLUTION

Let $A = B = x$. Then we can write

$$\cos(x + x) = \cos x \cos x - \sin x \sin x$$
$$\cos 2x = \cos^2 x - \sin^2 x.$$

EXAMPLE 4

Verifying a Trigonometric Identity

Use the trigonometric identity

$$\cos 2x = \cos^2 x - \sin^2 x$$

to verify that

$$\cos 2x = 2\cos^2 x - 1.$$

SOLUTION

$$\cos 2x = \cos^2 x - \sin^2 x$$
$$\cos 2x = \cos^2 x - 1 + \cos^2 x$$
$$\cos 2x = 2\cos^2 x - 1$$

SIMILAR PROBLEM

1. Given $\cos 2x = \cos^2 x - \sin^2 x$, obtain $\cos 2x = 1 - 2\sin^2 x$.

ANSWER

1. $\cos 2x = \cos^2 x - \sin^2 x$
$$= (1 - \sin^2 x) - \sin^2 x$$
$$= 1 - 2\sin^2 x$$

EXAMPLE 5 **Verifying a Trigonometric Identity**

Use the trigonometric identity

$$\cos 2x = 1 - 2\sin^2 x$$

to verify that

$$\sin^2 x = \frac{1 - \cos 2x}{2}.$$

SOLUTION

$$\cos 2x = 1 - 2\sin^2 x$$
$$2\sin^2 x = 1 - \cos 2x$$
$$\sin^2 x = \frac{1 - \cos 2x}{2}$$

SIMILAR PROBLEM

1. Given $\cos 2x = 2\cos^2 x - 1$, obtain $\cos^2 x = \frac{1}{2}(1 + \cos 2x)$.

ANSWER

1. $\cos 2x = 2\cos^2 x - 1$

$2\cos^2 x = 1 + \cos 2x$

$\cos^2 x = \dfrac{1}{2}(1 + \cos 2x)$

EXAMPLE 6 **Verifying a Trigonometric Identity**

Use the trigonometric identity

$$\sin(A + B) = \sin A \cos B + \cos A \sin B$$

to verify that

$$\sin 2x = 2\sin x \cos x.$$

SOLUTION

Let $A = B = x$, then we can write

$$\sin(x + x) = \sin x \cos x + \cos x \sin x$$
$$\sin 2x = 2\sin x \cos x.$$

EXAMPLE 7

Trigonometric Substitution

Write $\sqrt{x^2 + 4}$ in terms of θ when $x = 2 \tan \theta$.

SOLUTION

$$
\begin{aligned}
\sqrt{x^2 + 4} &= \sqrt{(2 \tan \theta)^2 + 4} \\
&= \sqrt{4(\tan \theta)^2 + 4} \\
&= \sqrt{4 \tan^2 \theta + 4} \\
&= \sqrt{4(\tan^2 \theta + 1)} \\
&= \sqrt{4} \sqrt{\tan^2 \theta + 1} \\
&= 2 \sqrt{\sec^2 \theta} \\
&= 2 \sec \theta, \qquad \sec \theta > 0
\end{aligned}
$$

SIMILAR PROBLEM

1. Evaluate $\sqrt{p^2 - 9}$ for $p = 3 \sec \theta$.

ANSWER

1. $3|\tan \theta|$

EXAMPLE 8

Converting the Product $\sin^n x \cos^m x$ to a Sum

Show that $\sin^5 x \cos^4 x = \sin^5 x - 2 \sin^7 x + \sin^9 x$.

SOLUTION

$$
\begin{aligned}
\sin^5 x \cos^4 x &= \sin^5 x (\cos^2 x)^2 \\
&= \sin^5 x [1 - \sin^2 x]^2 \\
&= \sin^5 x [1^2 - 2 \sin^2 x + (\sin^2 x)^2] \\
&= \sin^5 x [1 - 2 \sin^2 x + \sin^4 x] \\
&= \sin^5 x - 2 \sin^7 x + \sin^9 x
\end{aligned}
$$

SIMILAR PROBLEM

1. Show that $\cos^6 x \sin^3 x = \sin^3 x - 3\sin^5 x + 3\sin^7 x - \sin^9 x$.

ANSWER

1. $\cos^6 x \sin^3 x = (\cos^2 x)^3 \sin^3 x$

$$= (1 - \sin^2 x)^3 \sin^3 x$$
$$= (1 - 3\sin^2 x + 3\sin^4 x - \sin^6 x)\sin^3 x$$
$$= \sin^3 x - 3\sin^5 x + 3\sin^7 x - \sin^9 x$$

EXAMPLE 9

Converting $\sec^m x \tan^n x$ to a Sum

Show that $\sec^3 x \tan^5 x = \sec x \tan x(\sec^6 x - 2\sec^4 x + \sec^2 x)$.

SOLUTION

$$\sec^3 x \cdot \tan^5 x = (\sec x \sec^2 x) \cdot (\tan x \tan^4 x)$$
$$= \sec x \tan x \sec^2 x \tan^4 x$$
$$= \sec x \tan x \sec^2 x [\tan^2 x]^2$$
$$= \sec x \tan x \sec^2 x [\sec^2 x - 1]^2$$
$$= \sec x \tan x \sec^2 x [(\sec^2 x)^2 - 2\sec^2 x + 1]$$
$$= \sec x \tan x \sec^2 x [\sec^4 x - 2\sec^2 x + 1]$$
$$= \sec x \tan x [\sec^2 x \sec^4 x - \sec^2 x \cdot 2\sec^2 x + \sec^2 x]$$
$$= \sec x \tan x [\sec^6 x - 2\sec^4 x + \sec^2 x]$$

SIMILAR PROBLEM

1. Show that $\sec^5 x \tan^5 x = \sec x \tan x(\sec^8 x - 2\sec^6 x + \sec^4 x)$.

ANSWER

1. $\sec^5 x \tan^5 x = \sec x(\sec^4 x)[\tan x(\tan^2 x)^2]$

$$= \sec x \tan x \sec^4 x(-1 + \sec^2 x)^2$$
$$= \sec x \tan x \sec^4 x(1 - 2\sec^2 x + \sec^4 x)$$
$$= \sec x \tan x(\sec^8 x - 2\sec^6 x + \sec^4 x)$$

EXAMPLE 10

Reducing $\sin^n x$ to a Sum

Show that $\sin^4 x = \frac{3}{8} - \frac{1}{2}\cos 2x + \frac{1}{8}\cos 4x$.

SOLUTION

$$\sin^4 x = (\sin^2 x)^2$$
$$= \left[\frac{1}{2}(1 - \cos 2x)\right]^2$$
$$= \left[\frac{1}{2}\right]^2 (1 - \cos 2x)^2$$
$$= \frac{1}{4}[1 - 2\cos 2x + \cos^2 2x]$$

Now, since $\cos^2 2x = \frac{1}{2}[1 + \cos(2 \cdot 2x)] = \frac{1}{2} + \frac{1}{2}\cos 4x$, we have

$$\sin^4 x = \frac{1}{4}\left[1 - 2\cos 2x + \left(\frac{1}{2} + \frac{1}{2}\cos 4x\right)\right]$$
$$= \frac{1}{4}\left[\frac{3}{2} - 2\cos 2x + \frac{1}{2}\cos 4x\right]$$
$$= \frac{3}{8} - \frac{1}{2}\cos 2x + \frac{1}{8}\cos 4x.$$

SIMILAR PROBLEM

1. Show that $\cos^4 x = \frac{3}{8} + \frac{1}{2}\cos 2x + \frac{1}{8}\cos 4x$.

ANSWER

1. $\cos^4 x = (\cos^2 x)^2$
$$= \left(\frac{1 + \cos 2x}{2}\right)^2$$
$$= \frac{1}{4}(1 + 2\cos 2x + \cos^2 2x)$$
$$= \frac{1}{4}\left(1 + 2\cos 2x + \frac{1}{2}[1 + \cos 4x]\right)$$
$$= \frac{1}{4} + \frac{1}{2}\cos 2x + \frac{1}{8} + \frac{1}{8}\cos 4x$$
$$= \frac{3}{8} + \frac{1}{2}\cos 2x + \frac{1}{8}\cos 4x$$

EXAMPLE 11

A "Clever" Identity

Show that $\dfrac{1}{1 + \sin x} = \sec^2 x - \sec x \tan x.$

SOLUTION

The solution is based on the observation that

$$\cos^2 x = 1 - \sin^2 x = (1 + \sin x)(1 - \sin x). \qquad \text{Difference of squares}$$

Hence, we can write

$$\frac{1}{1 + \sin x} \cdot \frac{1 - \sin x}{1 - \sin x} = \frac{1 - \sin x}{1 - \sin^2 x}$$
$$= \frac{1 - \sin x}{\cos^2 x}$$
$$= \frac{1}{\cos^2 x} - \frac{\sin x}{\cos^2 x}$$
$$= \left(\frac{1}{\cos x}\right)^2 - \left(\frac{1}{\cos x}\right)\left(\frac{\sin x}{\cos x}\right)$$
$$= \sec^2 x - \sec x \tan x.$$

SIMILAR PROBLEM

1. Show that $\dfrac{1}{1 + \cos x} = \csc^2 x - \csc x \cot x.$

ANSWER

1. $\dfrac{1}{1 + \cos x} = \dfrac{1}{1 + \cos x} \cdot \dfrac{1 - \cos x}{1 - \cos x}$
$$= \frac{1 - \cos x}{1 - \cos^2 x}$$
$$= \frac{1 - \cos x}{\sin^2 x}$$
$$= \frac{1}{\sin^2 x} - \frac{\cos x}{\sin^2 x}$$
$$= \left(\frac{1}{\sin x}\right)^2 - \left(\frac{1}{\sin x}\right)\left(\frac{\cos x}{\sin x}\right)$$
$$= \csc^2 x - \csc x \cot x$$

EXAMPLE 12

Using the Relationships $x = r \cos \theta$ and $y = r \sin \theta$

Use the relationships $x = r \cos \theta$ and $y = r \sin \theta$ to show that $x^2 + y^2 = r^2$. Write the equation in terms of x, y, and r.

SOLUTION

$$x^2 + y^2 = (r\cos\theta)^2 + (r\sin\theta)^2$$
$$= r^2\cos^2\theta + r^2\sin^2\theta$$
$$= r^2(\cos^2\theta + \sin^2\theta)$$
$$= r^2 \cdot 1$$
$$= r^2$$

EXAMPLE 13

Using Relationships $x = r\cos\theta$ and $y = r\sin\theta$

Use the relationships $x = r\cos\theta$ and $y = r\sin\theta$ to write the equation $x^2 - y^2 = 1$ in terms of r and θ.

SOLUTION

$$x^2 - y^2 = 1$$
$$(r\cos\theta)^2 - (r\sin\theta)^2 = 1 \qquad \text{Substitution}$$
$$r^2\cos^2\theta - r^2\sin^2\theta = 1$$
$$r^2(\cos^2\theta - \sin^2\theta) = 1 \qquad \text{Factor out } r^2$$
$$r^2 \cdot \cos 2\theta = 1$$

EXAMPLE 14

Using the Relationships $x = r\cos\theta$ and $y = r\sin\theta$

Use the relationships $x = r\cos\theta$ and $y = \sin\theta$ to write the equation $r = 4/(1 - \cos\theta)$ in terms of x and y.

SOLUTION

Since θ can be substituted easily only when it occurs in the expressions $r\cos\theta$ or $r\sin\theta$, we convert $r = 4/(1 - \cos\theta)$ to a form that involves $r\cos\theta$ by multiplying through by $1 - \cos\theta$.

$$r(1 - \cos\theta) = 4$$
$$r - r\cos\theta = 4$$
$$r - x = 4$$
$$r = x + 4$$
$$r^2 = (x + 4)^2 \qquad \text{Square both sides}$$
$$x^2 + y^2 = x^2 + 8x + 16 \qquad \text{Substitution}$$
$$y^2 - 8x - 16 = 0$$

SIMILAR PROBLEM

1. Using the relationships $x = r\cos\theta$ and $y = r\sin\theta$, express the equation $r = 3/(1 + \sin\theta)$ in terms of x and y.

ANSWER

1. $x^2 + 6y - 9 = 0$

CHAPTER 15 EXERCISES

In Exercises 1 and 2, convert the expression to a simple trigonometric form.

1. $\sqrt{9 - x^2}$ when $x = 3\sin\theta$.   **2.** $\dfrac{1}{\sqrt{4 - x^2}}$ when $x = 2\sin\theta$.

In Exercises 3–6, $x = r\cos\theta$ and $y = r\sin\theta$. Express the equations in simple form in terms of r and θ. (Note that $x^2 + y^2 = r^2$.)

3. $x = y^2$ **4.** $(x + y)^2 - x + y = 0$

5. $x^2 + y^2 + 3x = 0$ **6.** $y = \dfrac{2x}{1 - x}$

7. Write $\cos^3 x \sin^4 x$ in terms of $\cos x$ (and its powers).

8. Show that $\tan^7 x \sec^5 x = \sec x \tan x (\tan^6 x + 2\tan^8 x + \tan^{10} x)$.

9. Write the expression in terms of $\sin\theta$ and $\cos\theta$.   **10.** Write the expression in terms of $\sin\theta$ and $\cos\theta$.

$$\frac{\csc\theta - \tan^2\theta}{\sec\theta - \cot\theta} \qquad\qquad \tan\theta - \frac{1 + \tan^2\theta}{\csc\theta + \cot\theta}$$

11. Express $\cos(2\arcsin x)$ without the use of trigonometric or inverse trigonometric functions.

12. Express $\sin(2\arctan x)$ without the use of trigonometric or inverse trigonometric functions.

13. Show that $\cos x \cdot \cos y = \frac{1}{2}[\cos(x + y) + \cos(x - y)]$.

14. Show that $\sin x \cdot \cos y = \frac{1}{2}[\sin(x + y) + \sin(x - y)]$.

In Exercises 15–18, assume $r\cos\theta = x$ and $r\sin\theta = y$. Express the equations in a simple form in terms of x and y.

15. $r = 3\cos\theta$ **16.** $r = \sin 2\theta$

17. $r = 4 - 3\sin\theta$ **18.** $r = \dfrac{1}{1 - 2\cos\theta}$

16

Sigma Notation, The Binomial Theorem, and Mathematical Induction

REVIEW OF FUNDAMENTALS

■ **Sigma Notation**

$$\sum_{i=p}^{n} a_i \text{ means } a_p + a_{p+1} + a_{p+2} + \cdots + a_n.$$

Example

$$\sum_{i=4}^{8} \frac{2i}{i-3} = \frac{2 \cdot 4}{4-3} + \frac{2 \cdot 5}{5-3} + \frac{2 \cdot 6}{6-3} + \frac{2 \cdot 7}{7-3} + \frac{2 \cdot 8}{8-3}$$

$$= \frac{8}{1} + \frac{10}{2} + \frac{12}{3} + \frac{14}{4} + \frac{16}{5}$$

$$= \frac{237}{10}$$

Properties

$$\sum_{i=p}^{n} (a_i + b_i) = \sum_{i=p}^{n} a_i + \sum_{i=p}^{n} b_i$$

$$\sum_{i=p}^{n} ca_i = c \cdot \sum_{i=p}^{n} a_i, \qquad \text{if } c \text{ does not depend on } i.$$

■ **Useful Formulas**

Formula	*Sigma Notation*
$\overbrace{c + c + \cdots + c}^{n \text{ terms}} = nc$	$\sum_{i=1}^{n} c = nc$
$1 + 2 + 3 + \cdots + n = \dfrac{n(n+1)}{2}$	$\sum_{i=1}^{n} i = \dfrac{n(n+1)}{2}$
$1^2 + 2^2 + \cdots + n^2 = \dfrac{n(n+1)(2n+1)}{6}$	$\sum_{i=1}^{n} i^2 = \dfrac{n(n+1)(2n+1)}{6}$
$1^3 + 2^3 + \cdots + n^3 = \dfrac{n^2(n+1)^2}{4}$	$\sum_{i=1}^{n} i^3 = \dfrac{n^2(n+1)^2}{4}$

■ **The Binomial Theorem**

$$(a+b)^n = a^n + na^{n-1}b + \frac{n(n-1)}{2}a^{n-2}b^2 + \frac{n(n-1)(n-2)}{2 \cdot 3}a^{n-3}b^3 + \cdots + nab^{n-1} + b^n$$

■ **Proofs by Mathematical Induction** apply to statements that are meant to be proved valid for all natural numbers n. The method for these proofs is as follows.

1. Check that the statement is valid in the special instance when $n = 1$.

2. Assuming the validity of the statement for $n = k$, prove the validity of the statement for $n = k + 1$.

EXAMPLE 1   **Using Summation Formulas**

Evaluate the following expressions.

(a) $\displaystyle\sum_{i=2}^{5} \frac{i+1}{i-1}$ (b) $\displaystyle\sum_{i=1}^{n} 1$

(c) $\displaystyle\sum_{i=2}^{4} 7n$ (d) $\displaystyle\sum_{i=2}^{4} 7i$

SOLUTION

(a) $\displaystyle\sum_{i=2}^{5} \frac{i+1}{i-1} = \frac{2+1}{2-1} + \frac{3+1}{3-1} + \frac{4+1}{4-1} + \frac{5+1}{5-1}$

$$= \frac{3}{1} + \frac{4}{2} + \frac{5}{3} + \frac{6}{4}$$

$$= \frac{49}{6}$$

(b) $\displaystyle\sum_{i=1}^{n} 1 = \underbrace{1 + 1 + \cdots + 1}_{n \text{ terms}} = n$

(c) $\displaystyle\sum_{i=2}^{4} 7n = 7n + 7n + 7n = 21n$

(d) $\displaystyle\sum_{i=2}^{4} 7i = 7 \cdot 2 + 7 \cdot 3 + 7 \cdot 4$

$$= 14 + 21 + 28$$

$$= 63$$

SIMILAR PROBLEMS

Evaluate the expressions.

1. $\displaystyle\sum_{i=3}^{5} \frac{i-2}{i+2}$ 2. $\displaystyle\sum_{i=1}^{n} 2$

3. $\displaystyle\sum_{i=3}^{6} 5n$ 4. $\displaystyle\sum_{i=3}^{6} 5i$

ANSWERS

1. $\dfrac{101}{105}$ 2. $2n$

3. $20n$ 4. 90

EXAMPLE 2   **Using Summation Formulas**

Are the following expressions true or false?

(a) $\displaystyle\sum_{i=1}^{n}(2ni^2 + i) = \sum_{i=1}^{n}2ni^2 + \sum_{i=1}^{n}i$

(b) $\displaystyle\sum_{i=1}^{n}(i + 1) = \left[\sum_{i=1}^{n}i\right] + 1$

(c) $\displaystyle\sum_{i=1}^{n}3i^2 = 3\sum_{i=1}^{n}i^2$

(d) $\displaystyle\sum_{i=1}^{n}\frac{i^2}{n^2} = i^2 \cdot \sum_{i=1}^{n}\frac{1}{n^2}$

(e) $\displaystyle\sum_{i=1}^{n}\frac{i^2}{n^2} = \frac{1}{n^2} \cdot \sum_{i=1}^{n}i^2$

SOLUTION

(a) True

(b) False. We should have

$$\sum_{i=1}^{n}(i + 1) = \sum_{i=1}^{n}i + \sum_{i=1}^{n}1.$$

$\sum_{i=1}^{n}1$ does not generally mean "1", it means "$1 + 1 + \cdots + 1$, n terms" which equals n.

(c) True

(d) False; it is permissible to "remove from under the $\sum$ sign" only factors which do *not* involve the summation index i.

(e) True, because

$$\sum_{i=1}^{n}\frac{i^2}{n^2} = \sum_{i=1}^{n}\left[\frac{1}{n^2}\right]i^2$$

and the factor $1/(n^2)$ does not depend on i.

SIMILAR PROBLEMS

Are the following expressions true or false?

1. $\displaystyle\sum_{i=1}^{n}(3ni+i^2)=\sum_{i=1}^{n}3ni+\sum_{i=1}^{n}i^2$ **2.** $\displaystyle\sum_{i=1}^{n}(i^2+1)=\sum_{i=1}^{n}i^2+1$

3. $\displaystyle\sum_{i=1}^{n}5i^2=5\sum_{i=1}^{n}i^2$ **4.** $\displaystyle\sum_{i=1}^{n}\frac{i^3}{n}=i^3\sum_{i=1}^{n}\frac{1}{n}$

5. $\displaystyle\sum_{i=1}^{n}\frac{i^3}{n}=\frac{1}{n}\sum_{i=1}^{n}i^3.$

ANSWERS

1. True   **2.** False   **3.** True

4. False   **5.** True

EXAMPLE 3

Using Summation Formulas

Convert to a form without $\sum$.

(a) $\displaystyle\sum_{i=1}^{n}\frac{i^2}{n^3}$ **(b)** $\displaystyle\sum_{i=1}^{n}\left(\frac{i}{n^2}+\frac{3}{n}\right)$

SOLUTION

(a) $\displaystyle\sum_{i=1}^{n}\frac{i^2}{n^3}=\frac{1}{n^3}\sum_{i=1}^{n}i^2$

$$=\frac{1}{n^3}\cdot\frac{n(n+1)(2n+1)}{6}$$

$$=\frac{(n+1)(2n+1)}{6n^2}$$

(b) $\displaystyle\sum_{i=1}^{n}\left[\frac{i}{n^2}+\frac{3}{n}\right]=\sum_{i=1}^{n}\frac{i}{n^2}+\sum_{i=1}^{n}\frac{3}{n}$

$$=\frac{1}{n^2}\cdot\sum_{i=1}^{n}i+\frac{3}{n}\sum_{i=1}^{n}1$$

$$=\frac{1}{n^2}\cdot\frac{n(n+1)}{2}+\frac{3}{n}\cdot n$$

$$=\frac{n+1}{2n}+3$$

$$=\frac{(n+1)+6n}{2n}$$

$$=\frac{7n+1}{2n}$$

SIMILAR PROBLEMS ▪▪▪▪▪▪▪▪▪

Convert to a form without $\sum$.

1. $\displaystyle\sum_{i=1}^{n} \frac{i}{n^2}$

2. $\displaystyle\sum_{i=1}^{n} \left(\frac{i^2}{n} + \frac{2}{n} \right)$

ANSWERS

1. $\dfrac{n+1}{2n}$

2. $\dfrac{2n^2 + 3n + 13}{6}$

EXAMPLE 4

Using Mathematical Induction

Prove that $\displaystyle\sum_{i=1}^{n} i^3 = \frac{n^2(n+1)^2}{4}$.

SOLUTION

Since we are being asked to prove a property which is meant to be valid for all natural numbers ($n = 1$, $n = 2$, etc.), we can apply mathematical induction.

1. Check that the statement

$$\sum_{i=1}^{n} i^3 = \frac{n^2(n+1)^2}{4}$$

is valid for the special instance when $n = 1$. In other words, we must check that

$$\sum_{i=1}^{1} i^3 = \frac{1^2(1+1)^2}{4}.$$

This is easy to do.

2. Assume the validity of the formula for $n = k$.

$$\sum_{i=1}^{k} i^3 = \frac{k^2[k+1]^2}{4}$$

This statement is sometimes called "the inductive hypothesis".

3. Prove the validity of the formula for $n = k + 1$.

$$\sum_{i=1}^{k+1} i^3 = \frac{(k+1)^2[(k+1)+1]^2}{4}$$

We shall convert the left side of this formula into the right side, using the inductive hypothesis.

$$\sum_{i=1}^{k+1} i^3 = \sum_{i=1}^{k} i^3 + (k+1)^3$$

$$= \frac{k^2(k+1)^2}{4} + (k+1)^3$$

$$= (k+1)^2\left[\frac{k^2}{4} + (k+1)\right]$$

$$= (k+1)^2\left[\frac{k^2 + 4(k+1)}{4}\right]$$

$$= (k+1)^2\left[\frac{k^2 + 4k + 4}{4}\right]$$

$$= (k+1)^2\left[\frac{(k+2)^2}{4}\right]$$

$$= \frac{(k+1)^2[(k+1)+1]^2}{4}$$

CHAPTER 16 EXERCISES

1. Write the following without the $\sum$ notation in the simplest form.

(a) $\displaystyle\sum_{i=1}^{n} 3$ (b) $\displaystyle\sum_{i=1}^{n} 2n$

(c) $\displaystyle\sum_{i=2}^{4} \frac{i^2}{i-1}$ (d) $\displaystyle\sum_{i=2}^{n} 1$

2. Which of the following is $\displaystyle\sum_{i=1}^{n} 3ni$ equal to?

(a) $3\displaystyle\sum_{i=1}^{n} ni$ (b) $3n\displaystyle\sum_{i=1}^{n} i$

(c) $3i\displaystyle\sum_{i=1}^{n} n$

3. Convert to the simplest form without $\sum$ notation.

$$\sum_{i=1}^{n}\left[\frac{i^3}{n^2} + \frac{i}{n} + n + 1\right]$$

4. Prove that $\displaystyle\sum_{i=1}^{n} i^4 = \frac{n(n+1)(6n^3 + 9n^2 + n - 1)}{30}$.

5. Prove that $\displaystyle\sum_{i=1}^{n} \frac{1}{i(i+1)} = \frac{n}{n+1}$.   6. Prove that, for $r \neq 1$, $\displaystyle\sum_{i=0}^{n} r^i = \frac{1 - r^{n+1}}{1 - r}$.

Appendix: Pretests

Chapter 1 Pretest 134

Chapter 2 Pretest 135

Chapter 3 Pretest 136

Chapter 4 Pretest 137

Chapter 5 Pretest 138

Chapter 6 Pretest 139

Chapter 7 Pretest 141

Chapter 8 Pretest 142

Chapter 9 Pretest 143

Chapter 10 Pretest 144

Chapter 11 Pretest 145

Chapter 12 Pretest 146

Chapter 13 Pretest 147

Chapter 14 Pretest 148

Chapter 15 Pretest 149

Chapter 16 Pretest 150

CHAPTER 1 PRETEST

1. Expand the following powers of binomials.

    (a) $(5x + 2)^3$ **(b)** $(x^4 - 1)^2$

2. Substitute $x + h$ for x in the expression $f(x) = x^3 - 7x^2 + x + 1$.

3. Are the following statements true or false?

    (a) $\dfrac{2k}{2x + h} = \dfrac{k}{x + h}$ **(b)** $\dfrac{1}{p + q} = \dfrac{1}{p} + \dfrac{1}{q}$

    (c) $\dfrac{x + y}{2} = \dfrac{x}{2} + \dfrac{y}{2}$ **(d)** $3 \cdot \dfrac{a}{b} = \dfrac{a}{3b}$

    (e) $3 \cdot \dfrac{a}{b} = \dfrac{3a}{b}$ **(f)** $3 \cdot \dfrac{a}{b} = \dfrac{3a}{3b}$

    (g) $3 \cdot \dfrac{a + b}{c} = \dfrac{3a + b}{c}$

4. Simplify the following compound fractions.

    (a) $\dfrac{x/2}{x/4}$ **(b)** $\dfrac{\dfrac{h}{x + h}}{h}$

5. Express $\dfrac{1}{x} + 1$ as a single fraction.

6. Show that $\dfrac{\dfrac{1}{x + h} - \dfrac{1}{x}}{h} = \dfrac{-1}{x(x + h)}$.

7. Show that $\dfrac{\sqrt{x + h} - \sqrt{x}}{h} = \dfrac{1}{\sqrt{x + h} + \sqrt{x}}$.

CHAPTER 2 PRETEST

1. Factor the expression $x^3 + 1$.

2. Factor the trinomial $4x^2 - 21x - 18$.

3. Factor the trinomial $2x^2 + x - 3$.

4. Factor the polynomial $3x^2 + 6x^3 - 9x$.

5. Factor $(x+1)^3(4x-9) - (16x+9)(x+1)^2$.

6. Simplify the quotient $\dfrac{x(5x+1) - 3(x^2+1)}{(x-1)^2}$.

7. Simplify the quotient $\dfrac{(x+1)^3(4x-9) - (16x+9)(x+1)^2}{(x-6)(x+1)^3}$.

CHAPTER 3 PRETEST

1. Solve $xy' + y = 1 + y'$ for y'.

2. Solve $2xy^3 + 3x^2 y' y^2 + 4 = 3x^2 y + x^3 yy' + 5y'$ for y'.

3. Solve the quadratic equation $4x^2 - 21x - 18 = 0$.

4. Solve the following quadratic equations.

   (a) $x^2 - x - 5 = 0$ **(b)** $x^2 + x + 1 = 0$

5. Solve the equation $2x - 2 = 1 + x - 2x^2$.

6. Solve $7x^4 - 42x^2 - 35x = 0$ for x.

7. Solve $(x + 1)^3 (4x - 9) - (16x + 9)(x + 1)^2 = 0$ for x.

8. Solve $\dfrac{x - 1}{(x + 1)(3x - 5)\sqrt{x + 3}} = 0$ for x.

9. Solve $\dfrac{7x^2 + 5x}{x^2 + 1} - \dfrac{5x}{x^2 - 6} = 0$ for x.

10. Solve $(x + 1)^{-1/2} + \dfrac{(x - 1)(x + 1)^{1/2}}{\sqrt{x^3 + 1}} = 0$ for x.

CHAPTER 4 PRETEST

1. Find the slope of the straight line that contains the two points $(1, -3)$ and $(-2, 5)$.

2. Find the slope of all straight lines perpendicular to the line described in Problem 1.

3. Each of the four straight lines shown has a slope that is approximately equal to one of the following values: -50, 0, 0.1, -1, -0.1, $\frac{1}{4}$, 2, 3, and $-\frac{1}{4}$. Choose the correct value for each line.

(a)

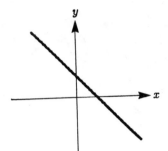

(b)

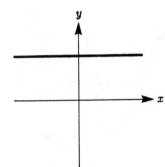

(c)

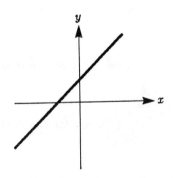

(d)

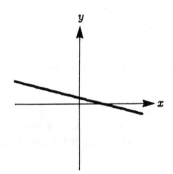

CHAPTER 5 PRETEST

1. The graph of the equation $y = x^3 - x$ is sketched in the accompanying figure. Answer the following questions precisely (i.e., do not estimate the answers from the graph).

   (a) Is the point $(3, 2)$ on the graph?

   (b) Is the point $(2, 6)$ on the graph?

   (c) What is the distance marked c?

   (d) What is the y-coordinate of P?

   (e) What are the coordinates of the points at which the curve intersects the x-axis?

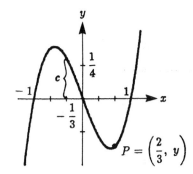

2. Find the equation of the straight line that passes through the point $(2, 4)$ and is parallel to the straight line $2x + 3y - 8 = 0$.

3. Find the equation of the straight line that is perpendicular to the line $2x + 3y - 8 = 0$ at the point $(1, 2)$.

4. The straight line with slope 5 that passes through the point $(-1, 3)$ intersects the x-axis at a certain point. What are the coordinates of this point?

5. What are the coordinates of the point at which the straight line passing through the points $(1, -3)$ and $(-2, -4)$ intersects the y-axis?

6. Sketch the graphs of $2x - 3y - 12 = 0$.   **7.** Sketch the graph of $x = 3$.

8. Sketch the graph of $y = 2x$.   **9.** Sketch the graph at $y = -x + 3$.

10. Sketch the graph of the equation $y = 2x^2 + x - 3$.   **11.** Sketch the graph of $y = -x^2 + 4x - 4$.

12. Sketch the graph of $y = -\dfrac{x^2}{3}$.   **13.** Sketch the graph of the equation $y = -x^2 - x - 1$.

14. Sketch the graph of $x^2 + y^2 = 3$.   **15.** Sketch the graph of $5x^2 + 2y^2 = 10$.

CHAPTER 6 PRETEST

1. Are the following statements true or false?

(a) $3 < 5$ (b) $3 \leq 5$ (c) $3 \leq 3$

(d) $5 < 1$ (e) $1 > 5$ (f) $5 > 1$

(g) $-2 < 5$ (h) $-7 < -2$

2. Are the following statements true for *all* values of x?

(a) $x^3 + 1 > x^3$ (b) $x^3 + x > x^3$ (c) $2x \geq x$

(d) $x^2 \geq 0$ (e) $x^2 \geq x$ (f) $\sqrt{x} \geq 0$

(g) $-x \leq 0$ (h) $\dfrac{1}{x} \leq x$

3. Solve the following inequalities for x.

(a) $x - 4 < 1$ (b) $3 - 2x \leq 1$ (c) $-1 < x - 4 < 1$

4. Solve the following inequalities for x.

(a) $2x^2 + x - 3 < 0$ (b) $2x^2 + x - 3 > 0$

5. Solve the inequality $2x^2 + 4x > 3x + 3$.

6. Solve the following inequalities.

(a) $4x^2 - 21x - 18 < 0$ (b) $4x^2 - 21x - 18 \leq 0$

7. Write each of the following in a form that does not contain an absolute value sign. The symbol x represents a real number and the symbol n represents a natural number (1, 2, 3, ..., and so on).

(a) $|1.35|$ (b) $\left|-\frac{1}{2}\right|$ (c) $|x|$

(d) $|x^2|$ (e) $|n|$ (f) $|n - 2|$

8. Fill in the blanks.

The statement $|x - 3| < 1$ is equivalent to $(\quad) < x < (\quad)$.

9. Show that if $|a - b| < s$ and $|c - b| < t$, then $|a - c| < s + t$.

10. Show that if $|x - 1| < \varepsilon/3$, then $|(3x + 2) - 5| < \varepsilon$.

11. Suppose that $|x - 3| < 1$. Show that the following statements are true.

(a) $|x| > 2$ **(b)** $|x + 3| < 7$

(c) If $|x - 3| < \dfrac{\varepsilon}{7}$ for some number ε, then $|x^2 - 9| < \varepsilon$.

(d) If $|x - 3| < 6\varepsilon$ for some number ε, then $\left| \dfrac{1}{x} - \dfrac{1}{3} \right| < \varepsilon$.

CHAPTER 7 PRETEST

1. Find the domains of the following functions.

   (a) $f(x) = x^{20}$

   (b) $g(x) = \dfrac{1}{x-3}$

   (c) $h(x) = \dfrac{1}{4x^2 - 21x - 18}$

   (d) $k(x) = \sqrt{4x^2 - 21x - 18}$

   (e) $p(x) = \dfrac{1}{\sqrt{4x^2 - 21x - 18}}$

2. Sketch the graph of $f(x) = 4x - 2$ and give its domain and range.

3. Sketch the graph of $g(x) = x$ and give its domain and range.

4. Sketch the graph of $h(x) = 3$ and give its domain and range.

5. Sketch the graph of $k(x) = c$ and give its domain and range.

6. Sketch the graph of $p(x) = 4 - x^2$ and give its domain and range.

7. Sketch the graph of $g(x) = 4 - x^2, \ \ 1 \le x < 3$ and give its domain and range.

8. For $g(x) = x^3 - 7x^2 + x + 1$, find $g(x + h)$.

9. For $k(x) = 3$, find $k(7)$ and $k(x + h)$.

10. For $f(x) = \dfrac{1}{x}$, find $\dfrac{f(x+h) - f(x)}{h}$.

11. Sketch the graph of $f(x) = x^2/4$. Let h be a positive number, and let x be in the domain of f. Indicate the following on the graph of f.

    (a) $f\left(\dfrac{3}{2}\right)$

    (b) $f\left(\dfrac{3}{2} + h\right)$

12. Sketch the graph of the function $f(x) = \begin{cases} 1, & \text{if } x \le 0 \\ -1, & \text{if } x > 0 \end{cases}$

13. Sketch the graph of the function $g(x) = \begin{cases} 2x, & \text{if } x < -1 \\ 2x^2 + x - 3, & \text{if } -1 \le x < 2 \\ -x + 3, & \text{if } x \ge 2 \end{cases}$

14. Sketch the graph of $f(x) = \sqrt{4 - x^2}$.

15. Obtain the formula for the function whose graph is the straight line passing through $(1, -3)$ and $(-2, -4)$.

CHAPTER 8 PRETEST

1. Let $f(x) = x^2 + 3$, $g(x) = 1/x$, and $k(x) = 2$. Find the following.

   (a) $(f + g)(x)$ **(b)** $(fg)(x)$ **(c)** $(k/f)(x)$

2. Let $f(x) = x^2 + 3$ and $k(x) = 2$. Express the function $q(x) = x^2 + 5$ as the sum of the two given functions.

3. Sketch the graph of the function $p(x) = (1/x) + 2$ by first sketching the graph of $g(x) = 1/x$.

4. For $f(x) = x^2 + 3$ and $g(x) = 1/x$, find the following.

   (a) $(f \circ g)(x)$ **(b)** $(g \circ f)(x)$

5. Express $h(x) = |2x - 4|$ as a composition of simpler functions.

6. Express the following functions as compositions of simpler functions.

   (a) $k(x) = \sqrt{3 - x}$ **(b)** $b(x) = \sqrt{x^2 - 2x}$

7. Graph the following functions.

   (a) $f(x) = |x|$ **(b)** $f(x) = |2x - 4|$

8. Let $f(x) = 2x - 2$.

   (a) Sketch the graph of f.
   (b) Determine whether f has an inverse function.
   (c) Sketch the graph of f^{-1}.
   (d) Give a formula for $f^{-1}(x)$.

10. Sketch the graph of $f(x) = \sqrt{x}$.

11. Sketch the graph of $g(x) = x^{1/3}$.

12. Graph the function $f(x) = 2x$. Find a number δ that satisfies the following condition.

    $$\text{Whenever } |x - 1| < \delta, \ |f(x) - 2| < 1.$$

13. Using the graph of $g(x) = \sqrt{x}$, find any number δ that satisfies the following condition.

    $$\text{Whenever } |x - 4| < \delta, \ |g(x) - 2| < 1.$$

CHAPTER 9 PRETEST

1. Solve the following simultaneous linear equations for x and y.

$$3x - 2y = 7, \qquad 4x + 5y = -6$$

2. Solve for x and y in the following pair of equations.

$$y - x + 1 = 0, \qquad 2y = 3x - 3x^2 + y^2$$

3. Solve for A, B, C, and D.

$$A + C = 0 \qquad \text{Equation 1}$$
$$B - D = 0 \qquad \text{Equation 2}$$
$$4A + 3D + 2C = 7 \qquad \text{Equation 3}$$
$$A + 4B + 6C = -6 \qquad \text{Equation 4}$$

4. (a) Find the points at which the straight lines $3x - 2y = 7$ and $4x + 5y = -6$ intersect.

(b) Find the x-coordinates of the points at which the curves $y - x + 1 = 0$ and $2y = 3x - 3x^2 + y^2$ intersect.

CHAPTER 10 PRETEST

1. Write $2x^2 + 7x + 11$ in the form $p[(x+q)^2 + r]$.

2. Write $\dfrac{-x^2}{3} + \dfrac{4x}{3} + 4$ in the form $s[(x+t)^2 - u^2]$.

3. Write the equation $16x^2 + 16y^2 + 56x - 64y - 31 = 0$ in the form $(x-h)^2 + (y-k)^2 = r^2$ where h, k, and r are constants. This is the standard form of the equation of a circle.

CHAPTER 11 PRETEST

1. Are the following statements true or false?

   (a) $(\sqrt{x})^2 = x$ for positive x

   (b) $\sqrt{x^2} = x$

   (c) $\sqrt{-x}$ is never a real number.

   (d) $\sqrt{x^2 + 1} = x + 1$

2. Write the following in the form x^n.

   (a) $\sqrt[3]{x}$

   (b) $\dfrac{1}{x}$

   (c) $\dfrac{1}{\sqrt{x}}$

   (d) $\dfrac{\sqrt[3]{x}}{x}$

3. Rewrite using fractional exponents.

   (a) $\sqrt{1 + x^2}$

   (b) $\dfrac{1}{\sqrt{(1 + z^2)^3}}$

4. Evaluate the exponential expression $(x^3 + 17)^{-3/4}$ when $x = 4$.

5. Show that $\dfrac{\sqrt{28}}{2} = \sqrt{7}$.

6. Express $\sqrt{4x^2 - 9}$ in the form $k\sqrt{x^2 - a^2}$.

7. Rewrite $\dfrac{u^2 - 3u + 2}{\sqrt{u}}$ as a sum using fractional exponents.

8. Sketch the graph of $f(x) = x^{2/3}$.

9. Sketch the graph of $g(x) = 2^x$ and give its domain and range.

10. Sketch the graph of $f(x) = a^x$, $a > 1$.

CHAPTER 12 PRETEST

1. Find the surface area of a box of height h whose base dimensions are p and q, and that satisfies either one of the following conditions.

   (a) The box is closed. (See figure.)

   (b) The box has an open top.

   (c) The box has an open top and a square base.

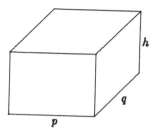

2. Find the surface area of a cylinder of radius r and height h and which satisfies either one of the following conditions.

   (a) The cylinder is closed.

   (b) The cylinder has an open top.

3. A seven-foot ladder, leaning against a wall, touches the wall x feet from the ground. Obtain an expression for the distance along the ground from the wall to the ladder.

4. A 6′ person is standing x feet away from a 10′ lamppost. What is the distance d from the base of the lamppost to the end of the person's shadow, expressed as a function of x?

5. A piece of wire 5 inches long is to be cut into two pieces. One piece is x inches long and is to be bent into the shape of a square. The other piece is to be bent into the shape of a circle. Find an expression for the total area made up by the square and the circle as a function of x.

CHAPTER 13 PRETEST

1. Evaluate $\log_9 3$.

2. Simplify the expression $\log_b b^{5x+1}$.

3. Simplify the expression $a^{\log_a x}$.

4. Sketch the graph of $f(x) = \log_b x$, and give its domain and range.

5. Using the fact that $\log_{10} 2 \approx 0.3010$, express $\log_{10}(x^2 + 1)$ in terms of $\log_2(x^2 + 1)$.

6. Write $\log_b \dfrac{(4x^5 - x - 1)\sqrt{x - 7}}{\sqrt[3]{x^2 + 1}}$ as a sum of logarithms.

7. Write $\log_a(3x + 2) - \dfrac{1}{3}\log_a(2x + 1) - 7\log_a(x^4 + x + 1)$ as a logarithm of a single expression.

CHAPTER 14 PRETEST

1. Evaluate the following.

   (a) $\cos 0$ **(b)** $\sin 0$ **(c)** $\tan \dfrac{\pi}{2}$ **(d)** $\cos \dfrac{\pi}{4}$

   (e) $\csc \dfrac{\pi}{3}$ **(f)** $\arccos \dfrac{\sqrt{3}}{2}$ **(g)** $\arctan 1$

2. Evaluate the following.

   (a) $\cos\left(-\dfrac{\pi}{4}\right)$ **(b)** $\sin \dfrac{4\pi}{3}$

   (c) $\cos 150°$ **(d)** $\tan \dfrac{2\pi}{3}$

3. For what values of θ is $\cos 4\theta = 0$?

4. If $f(x) = 2x - 1$ and $g(x) = \cos x$, find $(g \circ f)(x)$.

5. Express $h(x) = \sin(3x^2 + 5)$ as the composition of two simpler functions.

6. Which of the expressions are identical?

   (a) $\cos^2 x$ **(b)** $(\cos x)^2$

   (c) $\cos(x^2)$ **(d)** $(\sin x)^{-1}$

   (e) $\sin^{-1} x$ (meaning $\arcsin x$)   **(f)** $\sin(x^{-1})$

   (g) $\dfrac{1}{\sin x}$

7. Express $\sin(\arccos x)$ in a form that contains no trigonometric functions or their inverses.

CHAPTER 15 PRETEST

1. Use the trigonometric identity $\sin^2 x + \cos^2 x = 1$ to verify that $\tan^2 x + 1 = \sec^2 x$.

2. Use the identities $\cos(A - B) = \cos A \cos B + \sin A \sin B$ and $\cos(A + B) = \cos A \cos B - \sin A \sin B$ to verify that $\frac{1}{2}[\cos(A - B) + \cos(A + B)] = \cos A \cos B$.

3. Use the trigonometric identity $\cos(A + B) = \cos A \cos B - \sin A \sin B$ to verify that $\cos 2x = \cos^2 - \sin^2 x$.

4. Use the trigonometric identity $\cos 2x = \cos^2 x - \sin^2 x$ to verify that $\cos 2x = 2 \cos^2 x - 1$.

5. Use the trigonometric identity $\cos 2x = 1 - 2 \sin^2 x$ to verify that $\sin^2 x = \dfrac{1 - \cos 2x}{2}$.

6. Use the trigonometric identity $\sin(A + B) = \sin A \cos B + \cos A \sin B$ to verify that $\sin 2x = 2 \sin x \cos x$.

7. Write $\sqrt{x^2 + 4}$ in terms of θ when $x = 2 \tan \theta$.

8. Show that $\sin^5 x \cos^4 x = \sin^5 x - 2 \sin^7 x + \sin^9 x$.

9. Show that $\sec^3 x \tan^5 x = \sec x \tan x(\sec^6 x - 2 \sec^4 x + \sec^2 x)$.

10. Show that $\sin^4 x = \frac{3}{8} - \frac{1}{2} \cos 2x + \frac{1}{8} \cos 4x$.

11. Show that $\dfrac{1}{1 + \sin x} = \sec^2 x - \sec x \tan x$.

12. Use the relationships $x = r \cos \theta$ and $y = r \sin \theta$ to show that $x^2 + y^2 = r^2$. Write the equation in terms of x, y, and r.

13. Use the relationships $x = r \cos \theta$ and $y = r \sin \theta$ to write the equation $x^2 - y^2 = 1$ in terms of r and θ.

14. Use the relationships $x = r \cos \theta$ and $y = \sin \theta$ to write the equation $r = \dfrac{4}{1 - \cos \theta}$ in terms of x and y.

CHAPTER 16 PRETEST

1. Evaluate the following expressions.

   (a) $\displaystyle\sum_{i=2}^{5}\frac{i+1}{i-1}$
   (b) $\displaystyle\sum_{i=1}^{n}1$
   (c) $\displaystyle\sum_{i=2}^{4}7n$
   (d) $\displaystyle\sum_{i=2}^{4}7i$

2. Are the following expressions true or false?

   (a) $\displaystyle\sum_{i=1}^{n}(2ni^2+i)=\sum_{i=1}^{n}2ni^2+\sum_{i=1}^{n}i$
   (b) $\displaystyle\sum_{i=1}^{n}(i+1)=\left[\sum_{i=1}^{n}i\right]+1$

   (c) $\displaystyle\sum_{i=1}^{n}3i^2=3\sum_{i=1}^{n}i^2$
   (d) $\displaystyle\sum_{i=1}^{n}\frac{i^2}{n^2}=i^2\cdot\sum_{i=1}^{n}\frac{1}{n^2}$

   (e) $\displaystyle\sum_{i=1}^{n}\frac{i^2}{n^2}=\frac{1}{n^2}\cdot\sum_{i=1}^{n}i^2$

3. Convert to a form without $\sum$.

   (a) $\displaystyle\sum_{i=1}^{n}\frac{i^2}{n^3}$
   (b) $\displaystyle\sum_{i=1}^{n}\left(\frac{i}{n^2}+\frac{3}{n}\right)$

4. Prove that $\displaystyle\sum_{i=1}^{n}i^3=\frac{n^2(n+1)^2}{4}$.

Answers to Chapter Pretests

CHAPTER 1 PRETEST ANSWERS

1. **(a)** $125x^3 + 150x^2 + 60x + 8$
   (b) $x^8 - 2x^4 + 1$

2. $x^3 + 3x^2h + 3xh^2 + h^3 - 7x^2 - 14xh - 7h^2 + x + h + 1$

3. **(a)** False
   (b) False
   (c) True
   (d) False
   (e) True
   (f) False
   (g) False

4. **(a)** 2
   (b) $\dfrac{1}{x+h}$

5. $\dfrac{1+x}{x}$

CHAPTER 2 PRETEST ANSWERS

1. $(x+1)(x^2 - x + 1)$

2. $(4x+3)(x-6)$

3. $(2x+3)(x-1)$

4. $3x(2x+3)(x-1)$

5. $(x+1)^2(4x+3)(x-6)$

6. $\dfrac{2x+3}{x-1}$

7. $\dfrac{4x+3}{x+1}$

CHAPTER 3 PRETEST ANSWERS

1. $y' = \dfrac{1-y}{x-1}$

2. $y' = \dfrac{3x^2y - 2xy^3 - 4}{3x^2y^2 - x^3y - 5}$

3. $-\dfrac{3}{4}, 6$

4. **(a)** $\dfrac{1+\sqrt{21}}{2}, \dfrac{1-\sqrt{21}}{2}$
   (b) No real solution

5. $-\dfrac{3}{2}, 1$

6. $0, -1, \dfrac{1+\sqrt{21}}{2}, \dfrac{1-\sqrt{21}}{2}$

7. $-1, -\dfrac{3}{4}, 6$

8. 1

9. $0, -1, \dfrac{1+\sqrt{21}}{2}, \text{ and } \dfrac{1-\sqrt{21}}{2}$

10. 0

CHAPTER 4 PRETEST ANSWERS

1. $-\dfrac{8}{3}$

2. $\dfrac{3}{8}$

3. **(a)** -1
(b) 0
(c) 2
(d) $-\dfrac{1}{4}$

CHAPTER 5 PRETEST ANSWERS

1. **(a)** No
(b) Yes
(c) $8/27$
(d) $-10/27$
(e) $(-1,\ 0),\ (0,\ 0),\ (1,\ 0)$

2. $2x + 2y = 16$

3. $3x - 4y = -1$

4. $\left(-\dfrac{8}{5},\ 0\right)$

5. $\left(0,\ -\dfrac{10}{3}\right)$

6.

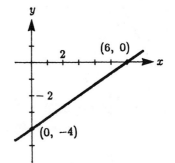

7.

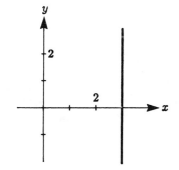

8.

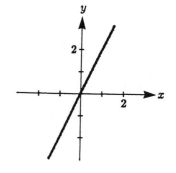

9.

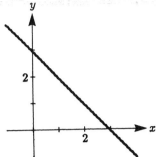

10.

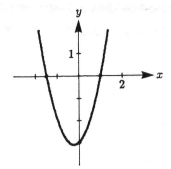

11.

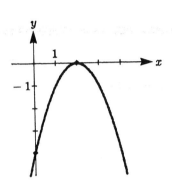

12.

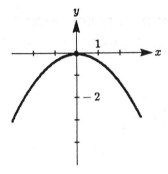

13.

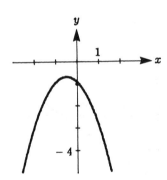

14.

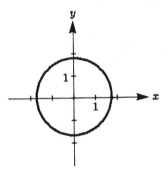

15.

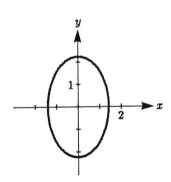

CHAPTER 6 PRETEST ANSWERS

1. **(a)** True   **(b)** True   **(c)** True
     (d) False   **(e)** False   **(f)** True
     (g) True   **(h)** True

2. **(a)** True   **(b)** False   **(c)** False
     (d) True   **(e)** False   **(f)** True
     (g) False   **(h)** False

3. **(a)** $x < 5$ **(b)** $x \geq 1$ **(c)** $3 < x < 5$

4. **(a)** $-\dfrac{3}{2} < x < 1$ **(b)** $\left(-\infty, \ -\dfrac{3}{2}\right), \ \ (1, \ \infty)$

5. $\left(-\infty, \ -\dfrac{3}{2}\right) \cup (1, \ \infty)$.   **6.** **(a)** $-\dfrac{3}{4} < x < 6$

                                                    (b) $-\dfrac{3}{4} \leq x \leq 6$

7. **(a)** 1.35 **(b)** $\frac{1}{2}$ **(c)** $|x| = \begin{cases} x, & \text{if } x \geq 0 \\ -x, & \text{if } x < 0 \end{cases}$

     (d) x^2 **(e)** n **(f)** $|n-2| = \begin{cases} 1 & \text{if } n = 1 \\ n - 2 & \text{if } n > 1 \end{cases}$

8. $2 < x < 4$

CHAPTER 7 PRETEST ANSWERS

1. **(a)** All real numbers
     (b) All real numbers except 3
     (c) All real numbers except $-3/4$ and 6
     (d) $(-\infty, \ -3/4]$ and $[6, \ \infty)$
     (e) $(-\infty, \ -3/4), \ \ (6, \ \infty)$.

2. Domain: all real numbers
     Range: all real numbers

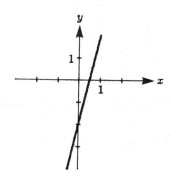

3. Domain: all real numbers
Range: all real numbers

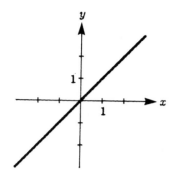

4. Domain: all real numbers
Range: 3

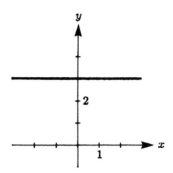

5. Domain: all real numbers
Range: c

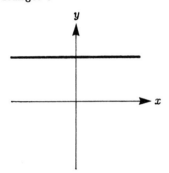

6. Domain: all real numbers
Range: $(-\infty, \ 4]$

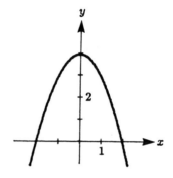

7. Domain: $[1, \ 3)$
Range: $(-5, \ 3]$

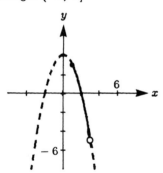

8. $x^3 + 3x^2h + 3xh^2 + h^3 - 7x^2 - 14xh - 7h^2 + x + h + 1$

9. $k(7) = 3$ and $k(x + h) = 3$

10. $\dfrac{-1}{(x + h)x}$

11.

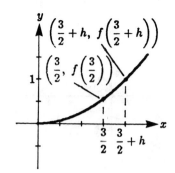

12.

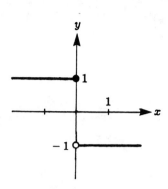

13.

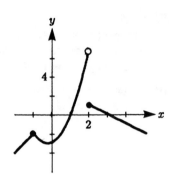

14.

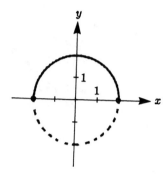

15. $f(x) = \dfrac{x}{3} - \dfrac{10}{3}$

CHAPTER 8 PRETEST ANSWERS

1.   (a) $x^2 + \dfrac{1}{x} + 3$

   (b) $x + \dfrac{3}{x}$

   (c) $\dfrac{2}{x^2 + 3}$

2. $q = f + k$

3.

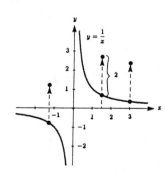

4.   (a) $\dfrac{1}{x^2} + 3$

   (b) $\dfrac{1}{x^2 + 3}$

5. Let $a(x) = |x|$ and $p(x) = 2x - 4$, then $h = a \circ p$.

6. (a) Let $s(x) = \sqrt{x}$ and $q(x) = 3 - x$, then $k = s \circ q$.

   (b) Let $s(x) = \sqrt{x}$ and $r(x) = x^2 - 2x$, then $b = s \circ r$.

7. **(a)**

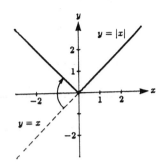

(b)

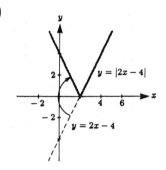

8. **(a)**

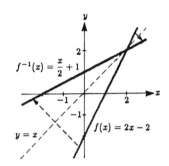

(b) Yes

(c) This graph is obtained by reflecting the original graph about the line $y = x$. A straight line is obtained.

(d) $f^{-1}(x) = \dfrac{x}{2} + 1$

9. **(a)**

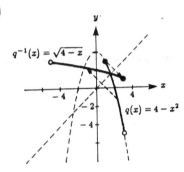

(b) Yes

(c) Reflection of original graph about the line $y = x$.

(d) $q^{-1}(x) = \sqrt{4 - x}, \quad -5 < x \le 3$

(e) Domain: $(-5, 3]$
Range: $[1, 3)$

10.

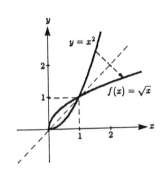

11.

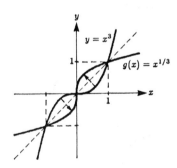

12.

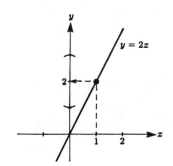 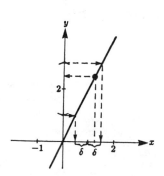

$\frac{1}{2}$ (Any positive value smaller than $\frac{1}{2}$ also would be acceptable.)

13.   3

CHAPTER 9 PRETEST ANSWERS

1. $x = 1,\ y = -2$

2. $x = -\frac{3}{2}$ and $y = -\frac{5}{2}$ or $x = 1$ and $y = 0$.

3. $A = 2,\ B = 1,\ C = -2,$ and $D = 1$.

4.   (a) $(1,\ -2)$
   (b) $(1,\ -\frac{3}{2})$

CHAPTER 10 PRETEST ANSWERS

1. $2\left[\left(x + \frac{7}{4}\right)^2 + \frac{39}{16}\right]$

2. $-\frac{1}{3}[(x - 2)^2 - 4^2]$

3. $\left(x - \left(-\frac{7}{4}\right)\right)^2 + (y - 2)^2 = 3^2$

CHAPTER 11 PRETEST ANSWERS

1.   (a) True
   (b) False
   (c) False
   (d) False

2.   (a) $x^{1/3}$
   (b) x^{-1}
   (c) $x^{-1/2}$
   (d) $x^{-2/3}$

3. (a) $\sqrt{1 + x^2} = (1 + x^2)^{1/2}$

   (b) $\dfrac{1}{\sqrt{(1 + z^2)^3}} = \dfrac{1}{[(1 + z^2)^3]^{1/2}} = \dfrac{1}{(1 + z^2)^{3/2}} = (1 + z^2)^{-3/2}$

4. $\dfrac{1}{27}$

6. $2\sqrt{x^2 - \left(\frac{3}{2}\right)^2}$

7. $u^{3/2} - 3u^{1/2} + 2u^{-1/2}$

8.

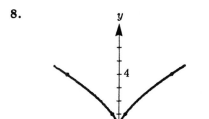

9. Domain: all real numbers
Range: $(0, \infty)$

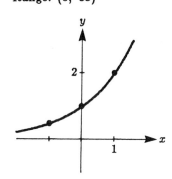

10.

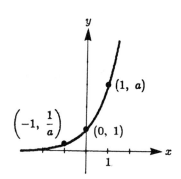

CHAPTER 12 PRETEST ANSWERS

1. **(a)** $2hp + 2hq + 2pq$
(b) $2hp + 2hq + pq$
(c) $4hp + p^2$

2. **(a)** $2\pi r^2 + 2\pi rh$
(b) $\pi r^2 + 2\pi rh$

3. $y = \sqrt{49 - x^2}$

4. $d = \dfrac{5x}{2}$

5. Total area $= \dfrac{x^2}{16} + \dfrac{25 - 10x + x^2}{4\pi}$

CHAPTER 13 PRETEST ANSWERS

1. $\dfrac{1}{2}$

2. $5x + 1$

3. x

4. Domain: all positive real numbers
Range: all real numbers

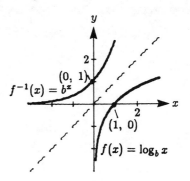

5. $0.3010 \log_2(x^2 + 1)$

6. $\log_b(4x^5 - x - 1) + \dfrac{1}{2}\log_b(x - 7) - \dfrac{1}{3}\log_b(x^2 + 1)$

7. $\log_a \dfrac{(3x + 2)}{\sqrt[3]{2x + 1}(x^4 + x + 1)^7}$

CHAPTER 14 PRETEST ANSWERS

1.
(a) 1
(b) 0
(c) undefined
(d) $\dfrac{\sqrt{2}}{2}$
(e) $\dfrac{2}{\sqrt{3}}$
(f) $\dfrac{\pi}{6}$
(g) $\dfrac{\pi}{4}$

2.
(a) $\dfrac{\sqrt{2}}{2}$
(b) $-\dfrac{\sqrt{3}}{2}$
(c) $-\dfrac{\sqrt{3}}{2}$
(d) $-\sqrt{3}$

3. $\dfrac{\pi}{8} + \dfrac{n\pi}{4}$

4. $\cos(2x - 1)$

5. $h = p \circ g$ where $p(x) = \sin x$, $g(x) = 3x^2 + 5$

6. Two identical pairs: **(a)** & **(b)**, **(d)** & **(g)**

7. $\sqrt{1 - x^2}$

CHAPTER 15 PRETEST ANSWERS

7. $2 \sec \theta, \qquad \sec \theta > 0$

13. $r^2 \cdot \cos 2\theta = 1$

14. $y^2 - 8x - 16 = 0$

CHAPTER 16 PRETEST ANSWERS

1.   (a) $\dfrac{49}{6}$

      (b) n

      (c) $21n$

      (d) 63

2.   (a) True

      (b) False

      (c) True

      (d) False

      (e) True

3.   (a) $\dfrac{(n+1)(2n+1)}{6n^2}$

      (b) $\dfrac{7n+1}{2n}$

Answers to Chapter Exercises

CHAPTER 1 EXERCISES

1. $x^3 + 3x^2h + 3xh^2 + h^3$

2. $x^6 - 9x^4 + 27x^2 - 27$

3. Not possible

4. $\dfrac{x}{y} + \dfrac{2}{3}$

5. $15x^2 + 15xh + 5h^2$

6. $\dfrac{3a + b^2}{b}$

7. $\dfrac{-a}{2x(2x + h)}$

8. $\dfrac{b(1 - b)}{b^2 + a}$

10. $\dfrac{3x^2 - 4x - 5}{(x - 1)^3}$

11. $\dfrac{2(x + 1)}{3x + 1}$

12. $\dfrac{-2x - h}{x^2(x + h)^2}$

CHAPTER 2 EXERCISES

1. $(y - 3)(y^2 + 3y + 9)$

2. $(x^2 + 4)(x + 2)(x - 2)$

3. $2(4x - 3)(10x + 9)$

4. $x^2(3x - 4)(x + 3)$

5. $x^2(x - 5)(4x + 1)$

6. $(x^2 + 2)(x^2 - 3)$

7. $(4x^2 - 1)(x^2 + 1)$

8. $2(2x - 1)^2(x - 1)(x + 2)$

9. $\dfrac{x + 2}{x - 2}$

10. $\dfrac{x}{2x + 1}$

11. $\dfrac{-(x + 1)^2}{2x(4x + 3)}$

12. $x - a = (\sqrt{x})^2 - (\sqrt{a})^2$
$$= (\sqrt{x} - \sqrt{a})(\sqrt{x} + \sqrt{a})$$

CHAPTER 3 EXERCISES

1. $\{2, -2\}$

2. 2

3. $\{2, -2\}$

4. $\dfrac{q + 1}{h - k - 6}$

5. $\left\{\dfrac{2}{3}, \dfrac{-3}{2}\right\}$

6. $\left\{\dfrac{1}{3}, 4\right\}$

7. 2

8. $\dfrac{7 \pm \sqrt{37}}{6} = \{2.18, \ 0.153\}$

9. $\dfrac{-3 \pm \sqrt{33}}{4} = \{-2.19, \ 0.69\}$

10. No real solution

11. No real solution.

12. $-\dfrac{5}{3}$

13. $t = \dfrac{5r}{d - 5}$

14. $\dfrac{xy + 3}{x + xy^2 - 5}$

15. $\dfrac{b}{t} = \dfrac{z-13}{r^2+s+2}$

16. $\dfrac{3x+2}{x-y^2-\dfrac{x^2y}{2}-2xy}$

17. $\dfrac{x-1}{-x^3+2x^2-x^2y+xy+1}$

18. $\left\{0,\ -2,\ \dfrac{1}{2}\right\}$

19. $\left\{-\dfrac{1}{2},\ -1,\ -2\right\}$

20. $\left\{-1,\ -\dfrac{1}{2},\ 1\right\}$

21. $\{-2,\ -1,\ 1\}$

22. $\{-3,\ -1,\ 0,\ 1\}$

23. $\left\{-1,\ -\dfrac{1}{2},\ 1,\ 5\right\}$

24. $1,\ \dfrac{-13\pm\sqrt{129}}{10}$

25. $\{0,\ 1\}$

26. $\left\{0,\ -1,\ \dfrac{3\pm\sqrt{13}}{2}\right\}$

27. $\{0,\ -2\pm3\sqrt{2}\}$

28. 5

29. 2

30. $\{1,\ -1\}$

31. 1

CHAPTER 4 EXERCISES

1. (b)

2. $\dfrac{q+1}{5-q}$

3. $-\dfrac{4}{3}$

4. (a) 2
   (b) 1
   (c) -1
   (d) 0
   (e) $\frac{1}{4}$

5. $-\dfrac{2}{3}$

CHAPTER 5 EXERCISES

1. (a) 3
   (b) 3
   (c) $-1,\ 1,\ \dfrac{3}{2}$

2.

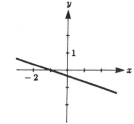

3.

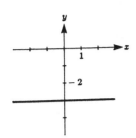

4.

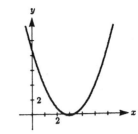

5.

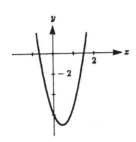

6.

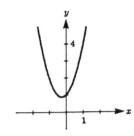

7.

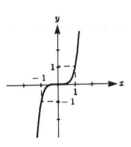

8.

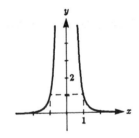

9.

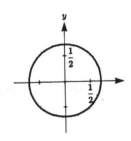

10.

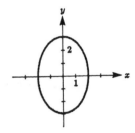

11. $2x + y + 11 = 0$

12. $3x - 4y - 15 = 0$

13. $x - 2y - 3 = 0$

14. $x + y - 2 = 0$

15. (a) $y = x$
(b) $y = 2x + 2$
(c) $y = 2$
(d) $x = 2$
(e) $3x + 2y - 6 = 0$

16. $x^2 + y^2 = 4$

17. $\dfrac{x^2}{9} + \dfrac{y^2}{4} = 1$

CHAPTER 6 EXERCISES

1. (a) T
   (b) T

2. (a) Yes
   (b) No; for example, not when $x = -1$.

3. (a) 0.01
   (b) $x^2 + 1$

4. (a) F
   (b) T
   (c) T

5. $x \leq -2$ or $x \geq 1$

6. $\frac{2}{3} \leq x \leq 4\frac{1}{2}$

9. 7

15. $x > 1$ and $-1 < x < 0$

16. $x > \frac{3}{2}$ and $-1 < x < 1$

17. $x \leq -2$ and $-\frac{1}{2} \leq x \leq 1$

CHAPTER 7 EXERCISES

1. Domain: All real numbers
   Range: All real numbers

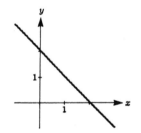

2. Domain: All real numbers
   Range: All real numbers; $x \geq 4$

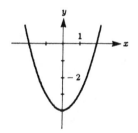

3. Domain: $(-2, -1]$ and $\left(\frac{1}{2}, 2\right)$
   Range: $\left(-\frac{3}{4}, 3\right)$

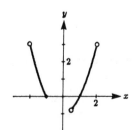

4.

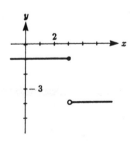

5. $-\dfrac{2x + h}{x^2(x + h)^2}$

6. -2

7. $\dfrac{x^3 + 3x^2h + 3xh^2 + h^3 - 2x - 2h}{x + h + 1}$

8. All real numbers except -5

9. All real numbers

10. All real numbers except 1 and -2

11. $-2 \leq x \leq 1$

12. $f(x) = -2x - 3$

13. $f(x) = \frac{1}{2}(x-3)$

14.

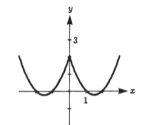

15. 4

16. -2

17. No; whatever $f(x)$ is chosen with $1 < x < 3$, it will be exceeded by the value of $f(x)$ for a slightly larger value of x less than 3.

18.

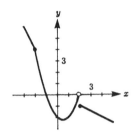

19.

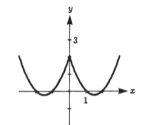

20. $-2 \le x \le 1$

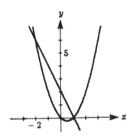

21. $f(x) = 2x, \quad g(x) = x$

CHAPTER 8 EXERCISES

1.

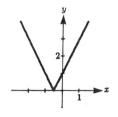

2. (a) $\dfrac{\sqrt{x^2+1}+1}{x}$

(b) $\dfrac{\sqrt{x^2+1}}{x}$

(c) $\sqrt{(1/x^2)+1}$

(d) $\dfrac{1}{\sqrt{x^2+1}}$

3. (a) $h \circ b$ where $h(x) = |x|$ and $b(x) = x^2 - 1$
(b) $h \circ b$ where $h(x) = \sqrt{x}$ and $b(x) = x^3 - 1$
(c) $h \circ b$ where $h(x) = x^{3/4}$ and $b(x) = 4x - 5$

4. $h \circ b$ where $h(x) = \dfrac{5}{\sqrt{x}} - 3x^2$

and $b(x) = 2x + 1$

5.

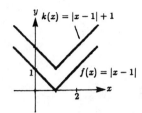

6. (a)

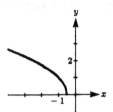

(b)

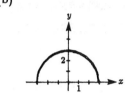

(c)

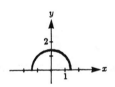

7. $f^{-1}(x) = \dfrac{2}{3} - \dfrac{x}{3}$

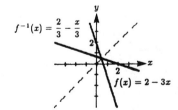

8.

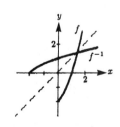

9. (a)

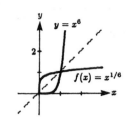

(b)

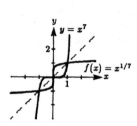

10. No; not identical for $f(x) = 2x$, for example

CHAPTER 9 EXERCISES

1. $x = 1$, $y = 2$

2. $(1, 2)$

3. $x = 2$ and $y = 3$, $x = 16$ and $y = -25$

4. $(-1, 2)$

5. $(2, 1)$, $(2, -1)$, $(-2, 1)$, $(-2, -1)$

6. $A = -6$, $B = 6$, $C = \dfrac{11}{2}$

7. $1, 3$

8. $A = -\dfrac{1}{2}$, $B = 3$, $C = \dfrac{7}{2}$, $D = -2$

9. $A = 2$, $B = -1$, $C = 1$, $D = -3$

CHAPTER 10 EXERCISES

1. $2\left[\left(x - \dfrac{3}{4}\right)^2 - \dfrac{1}{16}\right]$

2. $4\left[\left(x - \dfrac{1}{8}\right)^2 - \dfrac{17}{64}\right]$

3. $-2\left[\left(x + \dfrac{3}{4}\right)^2 - \dfrac{33}{16}\right]$

4. $3\left[\left(x - \dfrac{5}{6}\right)^2 - \dfrac{25}{36}\right]$

5. $3\left[\left(x + \dfrac{1}{6}\right)^2 - \dfrac{1}{36}\right]$

6. $-2[(x + 4)^2 - 5^2]$

7. $\dfrac{-3}{5}\left[\left(x - \dfrac{1}{4}\right)^2 - \left(\dfrac{3}{4}\right)^2\right]$

8. $\left(x + \dfrac{1}{4}\right)^2 + (y - 2)^2 = \left(\dfrac{7}{4}\right)^2$

9. $\left(x - \dfrac{1}{2}\right)^2 + \left(y + \dfrac{5}{6}\right)^2 = \left(\dfrac{1}{3}\right)^2$

10. $\dfrac{[x - (1/2)]^2}{(\sqrt{5}/2)^2} + \dfrac{[y - (-3/2)]^2}{(\sqrt{5})^2} = 1$

CHAPTER 11 EXERCISES

1. $(3x)^{1/5}$

2. $x^{-1/2}$

3. $x^{1/6}$

4. $(9 + x^2)^{-1/3}$

5. $(x^{1/2} + 1)^{-1/2}$

6. $\dfrac{1}{9}$

7. $\dfrac{1}{81}$

8. $\dfrac{1}{8}$

9. 4

10. $\dfrac{1}{4}$

11. $3\sqrt{2}$

14. $\sqrt{2}$

16. $3\sqrt{x^2 + 2^2}$

18. $\dfrac{1}{2}\sqrt{x^2 + 4^2}$

21. $\sqrt{23}\sqrt{x^2 - (\sqrt{5/23})^2}$

23. $u^{2/3} + u^{-1/3}$

24. $2u^{7/2} + 3u^{3/2} - u^{-1/2}$

25. $1 + u^{-1/6} + u^{-1/2}$

26. $5x^{5/2} - 2 - 4x^{-1/2}$

27.

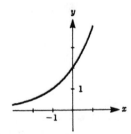

28.

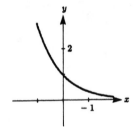

29.

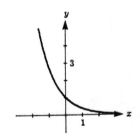

30.

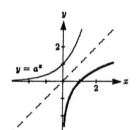

CHAPTER 12 EXERCISES

1. $wh - 3w - 2h + 6$ sq. in.

2. $wh + \dfrac{\pi w^2}{8}$

3. $\sqrt{25 + x^2}$

4. $4x^3 - 32x^2 + 63x$ cubic inches

5. $\pi q^2 h - \dfrac{\pi h^3}{4}$

6. $\dfrac{t^2}{16} + \dfrac{\sqrt{3}}{4}\left(9 - 2t + \dfrac{t^2}{9}\right)$

7. $\dfrac{\sqrt{3}t^2}{36} + \dfrac{64 - 16t + t^2}{4\pi}$

8. $t = \dfrac{3x}{7}$

9. $s = \dfrac{5r}{9 - r}$

10. $y = \dfrac{(x + 2)\sqrt{x^2 + 49}}{x}$

CHAPTER 13 EXERCISES

1. 0

2. $\dfrac{1}{2}$

3. $(0.6990) \log_5 x$

4. $\log_{10} p[\log_p(2x + 1)]$

5. 2

6. -2

7. $\log_b(2x^2 - 3) + 2\log_b(x^3 + 1) - \dfrac{1}{2}\log_b(7x + 1)$

8. $\log_b \dfrac{\sqrt{3x + 1}(1 - x)}{(1 - 9x^2)^{2/3}}$

9. $\log_b \dfrac{2\sqrt{x + 2}}{9(x - 2)^{5/2}}$

10.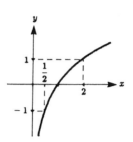

11.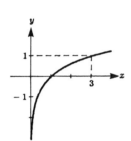

12. On the one hand, $\log_2(4 + 4) = \log_2 8 = 3$.
On the other hand, $\log_2 4 + \log_2 4 = 2 + 2 = 4$.

CHAPTER 14 EXERCISES

1. (a) $\dfrac{1}{2}$

   (b) $\sqrt{2}$

   (c) $\dfrac{1}{\sqrt{3}}$

2. (a) $-\dfrac{\sqrt{3}}{2}$

   (b) $\dfrac{\sqrt{3}}{2}$

   (c) $-\dfrac{\sqrt{3}}{2}$

   (d) $\dfrac{1}{2}$

   (e) $-\dfrac{1}{\sqrt{2}}$

3. (a) $\dfrac{1}{2}$

    (b) $\dfrac{1}{2}$

    (c) $\dfrac{1}{2}$

    (d) 0

4. (a) $\dfrac{\pi}{3}$

    (b) π

5. $\dfrac{\pi}{12} + \dfrac{k\pi}{6}, \quad k = \pm 1, \ \pm 2, \ \ldots$

6. $f \circ g$, where (a) $f(x) = \sin x$ and $g(x) = 2x - 1$, and (b) $f(x) = \arccos x$ and $g(x) = x^2 - 1$

7. $\dfrac{\sqrt{-x^2 - x}}{x + 1}$

8. Domain: all real numbers
Range: $[-1, \ 1]$

9. Domain: all real numbers except $\pm\dfrac{\pi}{2}, \ \pm\dfrac{3\pi}{2}, \ \ldots$
Range: all real numbers

CHAPTER 15 EXERCISES

1. $3\cos\theta$

2. $\dfrac{1}{2}\csc\theta$

3. $r = \csc\theta\cot\theta$

4. $r = \dfrac{\cos\theta - \sin\theta}{1 + 2\sin\theta\cos\theta}$

5. $r = -3\cos\theta$

6. $r = \dfrac{\sin\theta - 2\cos\theta}{\sin\theta\cos\theta}$

7. $\cos^3 x - 2\cos^5 x + \cos^7 x$

9. $\dfrac{\cos^2\theta - \sin^3\theta}{\cos\theta(\sin\theta - \cos^2\theta)}$

10. $\dfrac{\sin\theta\cos\theta(1 + \cos\theta) - \sin\theta}{\cos^2\theta(1 + \cos\theta)}$

11. $1 - 2x^2$

12. $\dfrac{2x}{1 + x^2}$

15. $x^2 + y^2 - 3x = 0$

16. $(x^2 + y^2)^3 - 4x^2y^2 = 0$

17. $16(x^2 + y^2) = (x^2 + y^2 + 3y)^2$

18. $3x^2 - y^2 + 4x + 1 = 0$

CHAPTER 16 EXERCISES

1. (a) $3n$
    (b) $2n^2$
    (c) $13\frac{5}{6}$
    (d) $n - 1$

2. (a) and (b)

3. $\dfrac{5n^2 + 8n + 3}{4}$

Index

Absolute values, 45, 51–55, 76
Algebra, elementary, 1–7
Arccos, arcsin, etc., 110–116
Area, 96–98, 101
Base of logarithm, 104, 106
Binomial Theorem, 127
Binomials, 2
Box, 97
Cancel (fractions), 3, 4
Circle,
    area of, 96, 101
    circumference of, 96, 98, 101
    equation of, 30, 41, 42
    graph of, 30, 42, 64
Completing the square, 84–87
Compound fractions, 4
Composition of functions, 62, 67, 69, 113
Cone, 96
Cylinder, 96, 98
Difference of cubes, 8
Difference of squares, 8
Domain of a function, 56–60, 73, 74, 93, 105
Ellipse, 30, 42
Equations, 13–23
    linear (solving), 13–15
    of circle, 30, 41
    of ellipse, 30, 42
    of straight lines, 30, 33
    polynomial (solving), 13, 18, 19
    quadratic (solving), 13, 15–17
Exponents, 88–93, 114
Factoring, 8–12
Fractions, 1, 3–6, 11
Functions, 56–77, 93, 94, 105
    composition of, 62, 67, 69, 113
    domain and range of, 56–60, 73, 74, 93, 105
    inverse, 67, 72–74
    products of, 67
    sums of, 67, 68
Graphs
    common, 30–43
    of functions, 56–59, 63–65, 68, 71–75, 93, 94, 105, 110
Highest common factor (HCF), 8–10
Identities, trigonometric, 117–126
Induction, mathematical, 127, 131
Inequalities, 45–55, 76
Intersecting curves, 79, 82

Interval notation, 46

Inverse functions, 67, 72–74

Inverse trigonometric functions, 110, 111, 115, 116

Lengths, formulas for, 96, 99, 100

Linear inequalities, 48

Logarithms, 104–107

Lowest common denominator (LCD), 1, 13, 21, 22

Mathematical induction, 127, 131

Normal lines, 25, 26, 33

Parabolas, 30, 38–41

Parallel lines, 33

Perpendicular lines, 25, 26, 33

Point-slope form, 30, 33

Polynomial(s),

    equations, 13, 18, 19

    factoring of, 8, 10, 11

Products of functions, 67

Pythagorean theorem, 96, 99

Quadratic inequalities, 50

Radian measure, 109, 111, 112

Radicals, 22, 46–48, 56, 57, 74, 88–92, 121

Range of a function, 56–59, 73, 74, 93, 105

Rationalizing numerators, 7

Reference angle, 110, 112

Set notation, 45

Sigma notation, 127–132

Similar triangles, 96, 100

Simultaneous equations, 79–82

Slope, 25–28, 30, 37

Slope-intercept form, 30, 33, 37

Sphere, 96

Square roots, 22, 47, 56, 57, 88–92, 121

Straight lines, sketching, 30, 35–37

Substitution,

   method,

     simultaneous equations, 79, 81, 82

   trigonometric, 121

Sum of cubes, 8

Sums of functions, 67, 68

Synthetic division, 18, 19

Trapezoid, 96

Triangle inequality, 46, 52

Trigonometry, 109–126

Trinomials, factoring, 9, 10

Two-point form of straight line, 30, 33, 35

Volume, 96